Active Networks and Active Network Management

A Proactive Management Framework

NETWORK AND SYSTEMS MANAGEMENT

Series Editor: Manu Malek
Lucent Technologies, Bell Laboratories
Holmdel, New Jersey

ACTIVE NETWORKS AND ACTIVE NETWORK MANAGEMENT:
A Proactive Management Framework

Stephen F. Bush and Amit B. Kulkarni

BASIC CONCEPTS FOR MANAGING TELECOMMUNICATIONS
NETWORKS: Copper to Sand to Glass to Air

Lawrence Bernstein and C. M. Yuhas

COOPERATIVE MANAGEMENT OF ENTERPRISE NETWORKS

Pradeep Ray

A Continuation Order Plan is available for this series. A continuation order will bring delivery of each new volume immediately upon publication. Volumes are billed only upon actual shipment. For further information please contact the publisher.

Active Networks and Active Network Management
A Proactive Management Framework

Stephen F. Bush

and

Amit B. Kulkarni
General Electric Corporate Research and Development
Niskayuna, New York

KLUWER ACADEMIC/PLENUM PUBLISHERS
New York, Boston, Dordrecht, London, Moscow

Library of Congress Cataloging-in-Publication Data

Bush, Stephen F.
   Active networks and active network management: a proactive management framework
/Stephen F. Bush, Amit B. Kulkarni
     p.   cm.—(Network and systems management)
   Includes bibliographical references and index.
   ISBN 0-306-46560-4
     1. Computer networks—Management. I. Kulkarni, Amit B.   II. Title. III. Kluwer
Academic/Plenum Publishers network and systems management

TK5105.5 .B877 2001
004.6—dc21

00-054581

ISBN 0-306-46560-4

©2001 Kluwer Academic/Plenum Publishers, New York
233 Spring Street, New York, N.Y. 10013

http://www.wkap.nl/

10 9 8 7 6 5 4 3 2 1

A C.I.P. record for this book is available from the Library of Congress

Printed in the United States of America

This book is dedicated to Ann Bush and Asavari Kulkarni for their toleration of the long hours spent on this effort. It is also dedicated to Stephen's parents and Amit's parents, who are a source of inspiration for them. We would also like to recognize Mike Hartman, Don Puckette, and Kirby Vosburgh for their courage, support, and willingness to promote and support new ideas.

FOREWORD

Steven F. Bush and Amit B. Kulkarni

This book is designed to meet the needs of business and industry, new researchers, and as a supplement to graduate study. The first chapter provides an overview of the benefits and applications of active networks to business and industry. The remainder of the book is designed to bring someone familiar with telecommunications up to date on the concepts behind active networks. Finally, student exercises are included within the text for those wishing to explore the topic in a more active manner.

The CD included with this book contains the full AVNMP version 1.1 source code required to enhance an application with predictive capability. AvnmpTutorial.htm is the top-level HTML file on the CD.

PREFACE

Active networking is an exciting new paradigm in digital networking that has the potential to revolutionize the manner in which communication takes place. It is an emerging technology, one in which new ideas are constantly being formulated and new topics of research are springing up even as this book is being written. This technology is very likely to appeal to a broad spectrum of users from academia and industry. Therefore, this book was written in a way that enables all these groups to understand the impact of active networking in their sphere of interest. Information services managers, network administrators, and e-commerce developers would like to know the potential benefits of the new technology to their businesses, networks, and applications. The book introduces the basic active networking paradigm and its potential impacts on the future of information handling in general and communications in particular. This is useful for forward-looking businesses that wish to actively participate in the development of active networks and ensure a head start in the integration of the technology in their future products, be they applications or networks. Areas in which active networking are likely to make significant impact are identified, and the reader is pointed to any related ongoing research efforts in the area.

The book also provides a deeper insight into the active networking model for students and researchers, who seek challenging topics that define or extend frontiers of the technology. It describes basic components of the model, explains some of the terms used by the active networking community, and provides the reader with taxonomy of the research being conducted at the time this book was written. Current efforts are classified based on typical research areas such as mobility, security, and management. The intent is to introduce the serious reader to the background regarding some of the models adopted by the community, outline to outstanding issues concerning active networking, and to provide a snapshot of the fast-changing landscape in active networking research.

Management is a very important issue in active networks because of its open nature. The latter half of the book explains the architectural concepts of a model for managing active networks and the motivation for a reference model that addresses limitations of the current network management framework by leveraging the powerful features of active networking to develop an integrated framework. It also describes a novel application enabled by active network technology called the Active Virtual Network Management Prediction (AVNMP) algorithm. AVNMP is a pro-active management system; in other words, it provides the ability to solve a potential problem before it impacts the system by modeling network devices within the network itself and running that model ahead of real time.

This book comes with a CD-ROM that includes a tutorial for those readers who like to take a hands-on approach to learning. The CD-ROM contains a fully functional version of the Active Virtual Network Management Prediction (AVNMP version 1.1) source code along with an HTML-based tutorial that can be easily viewed using any HTML browser. The tutorial instructs the user in a step-by-step fashion on how to enhance an active network application with AVNMP.

This book has benefited greatly from the review by a number of people, who generously gave us their time and expertise. We are thankful to the following people for providing us with invaluable comments to improve the book: Dan Raz, Lucent Technologies, Bell Laboratories; Rolf Stadler, Columbia University; and Eckhard Moeller, GMD FOKUS, Germany. The authors extend a special acknowledgment to Manu Malek for his patience and time in helping us shepherd this book into a more presentable form.

The authors would also like to acknowledge the funding provided by DARPA/ITO Contract Number: F30602-98-C-0230 supported by the Air Force Research Laboratory/IF that allowed us to go forward with research in this area.

CONTENTS

PART II: ACTIVE NETWORK ARCHITECTURE

PART III: AVNMP

Active Networks and Active Network Management

A Proactive Management Framework

1

INTRODUCTION

Active networking is a novel approach to network architecture in which network nodes – the switches, routers, hubs, bridges, gateways etc., – perform customized computation on the packets flowing through them. The network is called an "active network" because new computations are injected into the nodes dynamically, thereby altering the behavior of the network. Packets in an active network can carry fragments of program code in addition to data. Customized computation is embedded in the packet's code, which is executed on the network nodes. By making the computation application-specific, applications utilizing the network can customize network behavior to suit their requirements and needs.

The active network model provides a user-driven customization of the infrastructure, allowing new services to be deployed at a faster pace than can be sustained by vendor-driven consensus or through standardization. The essential feature of active networks is the programmability of its infrastructure. New capabilities and services can be added to the networking infrastructure on demand. This creates a versatile network that can easily adapt to future needs of applications. The ability to program new services into the network will lead to a user-driven innovation process in which the availability of the new services will be dependent on their acceptance in the marketplace. In short, active networking enables the rapid deployment of novel and innovative services and protocols into the network. For example, a video conferencing application can inject a custom packet-filtering algorithm into the network that, in times of congestion, filters video packets and allows only audio packets to reach the receivers. Under severe congestion conditions, the algorithm compresses audio packets to reduce network load and alleviate congestion. This enables the application to handle performance degradation due to network problems gracefully and in an application-specific manner.

In active networking, applications cannot only determine the protocol functions as necessary at the endpoints of a communication path, but can also inject new protocols into the network for the network nodes to execute on their behalf. The nodes of the network, called active nodes, are programmable entities. Application code executes on these nodes to implement new protocols and services.

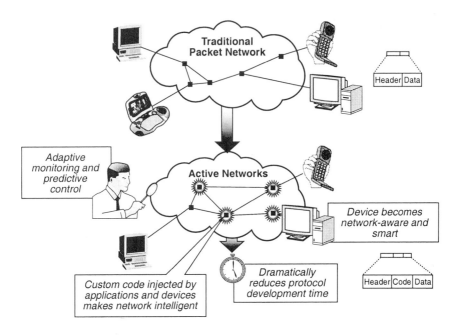

Figure 1.1 Active Network Overview.

Active networking is a radical new paradigm for many in communication research. Researchers from all over the world are actively investigating this new technology. However, active networking is not confined only to academic research. High profile networking giants like Lucent, Nortel and Hitachi are actively participating in the effort to make these goals realizable. Active networks have caught the eye of conference organizers and journal publishers. This is evident from conferences like the International Workshop on Active Networks (IWAN) held in Germany in 1999, and the Intelligence in Services and Networks (IS&N) conference in Greece in 2000. Active networks are all-pervasive because they enhance all applications that need communications capabilities. That is why active networks are included in the list of topics at conferences on wireless as well as high-speed networks, signaling structures, distributed computing, and also real-time systems and applications.

1.1 BACKGROUND

The concept of active networking evolved from discussions within the Department of Defense's Defense Advanced Research Projects Agency (DARPA) community regarding future directions in communication networks. The purpose of these discussions was to design networks that overcome the problems and limitations of existing networks. Some of the problems identified were:

- Existing networks cannot support rapid development and deployment of new protocols and services. Designing a new protocol requires acceptance from a standardization committee and later implementation and support from vendors before applications can use the protocol. Experience has shown that this is usually a long and tedious process. For example, work on the design of the Internet Protocol version 6 (IPv6) was initially started in 1995, but the protocol has still not found widespread deployment.
- The existing Internet model allows limited security in the form of secure sockets, application data encryption and firewalls, but it is still difficult to create seamless, secure Virtual Private Networks over the Internet.
- Network functionality is scattered across routers, switches, end-systems, link-level hardware and applications, which prevents gathering of coherent state and makes network management extremely difficult.
- Mobile devices have to be treated as static devices because it is not currently possible to move a mobile state seamlessly across the network.
- It is not possible to deploy new services such as adaptive transcoding or in-network data-fusion, when and where you want them dynamically. In current networks, services are rigidly built into the kernels of operating systems of end-systems and network devices, preventing on-the-fly customization of these services by the applications.

To overcome the above limitations, DARPA began working on a set of ideas collectively referred to as active networking. The principal tenet of active networking is to devise a network architecture so that new features and services can be programmed into the network. The idea of carrying computation into the network is a step in the evolution from circuit switching and packet switching to an active network, in terms of increasingly greater computational role in the network, as shown in Figure 1.2.

When combined with well-defined execution environments at the network nodes, this program-based approach provides a foundation for expressing networking systems as a composition of smaller components with each component possessing specific properties. Network services are configured modularly and distributed within the network to meet application needs. Overall, the goals of active networking are:

- To create an architecture that enables solutions for immediate and future needs to be conceived and implemented easily.
- To provide a quantifiable improvement in the number and applicability of services that can be provided in the network.
- To enable application-specified control of network resources.
- To develop and deploy network security from the "ground up," i.e., to develop an infrastructure that is composed of secure components instead of adding security only at certain points in the network, such as firewalls and end-points of a connection path.

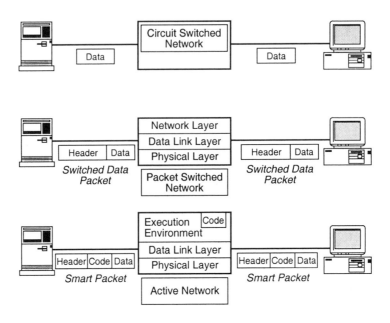

Figure 1.2. Network Evolution.

1.2 OUTLINE OF THE BOOK

Active networking is an emerging technology, one in which new ideas are constantly being formulated and new topics of research are springing up even as this book is being written. It is therefore impossible to cover the entire topic in a single publication. Instead, this portion of the book utilizes a different approach. The basis of the approach is that this technology will appeal to a broad spectrum of users from the academia to industry. However, people with different interests will expect to see some treatment of their interest in the book. Information Services managers, network administrators, e-commerce developers would like to know the potential benefits that the new technology brings to their businesses, networks and applications. Students expect to see a tutorial-like introduction to the concepts. On the other hand, researchers seek challenging topics that define or extend frontiers of the technology.

This book is divided in three parts to cater to the diverse needs of the target audience. In the first part, the reader is acquainted with the basic concepts of the technology. This part is intended for people who would like to quickly grasp the technology and evaluate its potential for their specific situation. The material is purposefully kept simple and high-level to enable technology-savvy managers and interested researchers to get a first glimpse of this new concept. Readers of this part will gain a basic knowledge of the active networking model, understand the motivation for developing such a model, and grasp its impacts on current applications and networks. Accordingly, the chapters in the first part of the book are organized as follows:

Part I: Introduction to Active Networking
- Chapter 1: Introduction
- Chapter 2: Properties of Active Networks

The first chapter introduces the basic active networking paradigm. Areas in which active networking is likely to make significant impact are identified, and the reader is pointed to any ongoing related research in these areas. This chapter is useful for users who wish to actively participate in the development of active networks and ensure a head start in the integration of the technology in their future products, be they applications or networks. Areas in which active networking is likely to have a significant impact are identified and the reader is pointed to any related ongoing research efforts in the area. Chapter 2 provides further insight into the model. It describes basic components of the model and explains some of the terms used by the active networking community. It also provides the reader with a taxonomy of the research being conducted at the time this book was written. Current efforts are classified based on typical research areas such as mobility, security and management. The intent here is to introduce the background of some of the models adopted by the community, to outline outstanding issues concerning active networking, and to provide a snapshot of the fast-changing landscape in active networking research. This chapter is particularly useful for those interested in pursuing new research in their area of interest as it relates to active networking.

The second part of the book provides a basis for understanding the architectural concepts of the model. Specifically, the architectural frameworks of an active network and a model for active network management are discussed. The chapters in this part are organized as follows:

Part II: Active Networking Architecture
- Chapter 3: Architectural Framework
- Chapter 4: Management Reference Model

Chapter 3 takes a look at the current version of the architectural specification of an active network. It discusses various aspects of the document and provides a background for some of the concepts outlined therein. Chapter 4 discusses the motivation for a reference model that addresses limitations of the current network management framework and leverages the powerful features of active networking to develop an integrated framework. The later part of Chapter 4 prepares the reader for AVNMP, which is the focus of the remainder of the book.

The final part of the book provides a close-up view of a novel application enabled by active network technology. It describes the life-cycle of an active networking protocol from conception to implementation. The application chosen implements the predictive aspect of the active management framework discussed in Chapter 4 and is called Active Virtual Network Management Prediction. In current network management, managed entities are either polled to determine their health or they send unsolicited messages indicating failed health. By the time such messages are generated, much less received, by a centralized system manager, the network has already failed. Active Virtual Network Management Prediction has resulted from research in developing pro-active system management, in other words, to solve a potential problem before it impacts the system. Active Virtual Network Management Prediction accomplishes this by modeling network devices within the network itself and running that model ahead of real time. Active Virtual Network Management Prediction is also self-correcting. Thus, managed devices can be queried for events which are likely to happen in the future; problems are detected ahead of time. The chapters in this part are organized as follows:

Part III: Active Virtual Network Management Protocol
- Chapter 5: AVNMP Architecture
- Chapter 6: Detailed Example of AVNMP Operation
- Chapter 7: Algorithmic Description of AVNMP
- Chapters 8-9: Performance Measurements and Analysis of AVNMP

Chapter 5 describes the architecture of the AVNMP framework and explains how various features of an active network can be leveraged to create a novel management strategy. A good time to begin using the CD tutorial is after reading Chapter 5 (AvnmpTutorial.htm). Chapter 5 includes examples of Driving Processes for specific applications, while Chapter 6 provides a detailed operational example of AVNMP. Chapter 7 discusses the background and origin of the algorithm used by AVNMP and includes an Appendix on some of the implementation details. Chapter 8 quantifies the performance of AVNMP, deriving equations for AVNMP performance and overhead. Chapter 9 completes the book by considering the challenges faced by any system attempting to predict its own behavior and some of the unique characteristics of AVNMP in meeting those challenges.

I

INTRODUCTION TO ACTIVE NETWORKING

2

PROPERTIES OF ACTIVE NETWORKS

Active networking provides a programmable infrastructure using a well-defined structure for packets that contain general-purpose program code and a uniform standardized execution platform at the nodes of the network. This chapter gives the reader an overview of the basic model of an active network. It describes basic components and introduces common terminology. It also describes the efforts of the active networking community to develop an active network backbone for collaborative research. Finally, it discusses a few popular execution environments. This discussion involves tools that were available in the spring of 2000 and it will be useful for someone evaluating active network prototypes for their own research and applications.

2.1 ACTIVE NETWORKING MODEL

Future applications will have to combat a myriad of network bandwidths ranging from a few kilobits per second to gigabits per second. Applications have to deliver information to users over a variety of access technologies from optic fiber to phone lines to wireless. Applications will need to be network-aware to adapt to the constraints of the underlying networking hardware. The current networking model has severe limitations that prevent it from meeting the demanding needs of emerging and future applications and allowing users "anytime, anywhere" access to information.

However, traditional networking protocols were built for largely non-real time data with very few burst requirements. The protocol stack at a network node is fixed and the network nodes only manipulate protocols up to the network layer. New protocols such as RTP and HTTP enable the network to transport other types of application data like real time and multimedia data. Such protocols cater to specific demands of the application data. Transporting those new data types over a legacy network requires the transformation of the data of a new type into data of a type carried by the network. However, transforming the data to fit legacy protocol requirements prevents one from understanding the transformed protocol. For example, embedding an MPEG frame in MIME format prevents one from easily recognizing an I, P or B frame. This prevents the network from taking suitable action on the MPEG frame during times of congestion. If information about the frame (e.g., the type of the frame, whether I, P or B) is not converted into MIME but the frame itself is converted, then both the goals of encoding and congestion control are satisfied.

Traditional protocol frameworks use layering as a composition mechanism. Protocols in one layer of the stack cannot guarantee anything about the properties of the layers underneath it. Each protocol layer is treated like a black box, and there is no mechanism to identify whether functional redundancies occur in the stack. Sometimes, protocols in different layers of the same stack need to share information. For example, TCP calculates a checksum over the TCP message and the IP header. But this action violates modularity of the layering model because the TCP module needs information from the IP header that it gets by directly accessing the IP header. Furthermore, layering hides functionality, which can introduce redundancy in the protocol stack. Introducing new protocols in the current infrastructure is a difficult and time-consuming process. A committee has to agree on the definition of a new protocol. This involves agreeing on a structure, states, algorithms, and functions for the protocol. All these issues require a consensus agreement on the part of a committee that standardizes the protocol. Experience has shown that the standardization is a time-consuming process. The time from conceptualization of a protocol to its actual deployment in the network is usually a matter of years. For example, work on the design of the Internet Protocol version 6 (IPv6) was initially started in 1995, but the protocol has still not found widespread deployment. Once the standardization process is completed, it is followed by identical implementations of the protocol in all devices. However, variations in the implementation by different network hardware vendors causes problems for interoperability. Vendor implementation of a protocol may differ if the vendors provide value-added features in their device or if they tweak the implementation to exploit hardware-specific features.

Another issue that vendors have to deal with is backward compatibility. A revision of a protocol may need to change the positions of the bits in the header of the protocol to accommodate more information. However, network devices upgraded with the new protocol still have to support data that conforms to the earlier revision. For example, the address field in the Internet Protocol (version 4) is 32 bits, as defined in the standards document. This implies that the protocol (and hence the network) supports a maximum of 2^{32} addresses. The tremendous growth of the Internet and the introduction of Internet-capable devices indicates that we are likely to run out of IP numbers in a very short time. Increasing the length of the address field in the IP header is not a solution to this problem because implementing the revised protocol is a formidable task. Increasing the length of the address field affects the positions of the succeeding fields in the protocol header as the bits are shifted by the appropriate number of positions. All software related to IP layer processing relies on the fields being in their correct positions, and therefore all existing communication software would have to be re-written to accommodate the change. This requires updating tens of thousands of existing routers and switches that implement IPv4 with the new protocol software.

Active networking provides a flexible, programmable model of networking that addresses the concerns and limitations described above. In an active networking paradigm, the nodes of the network provide execution environments that allow execution of code dynamically loaded over the network. Thus, instead of standardizing individual protocols, the nodes of the network present a standardized execution platform for the code-carrying packets. This approach eliminates the need for network-wide standardization of individual protocols. New protocols and services can be rapidly introduced in the network. Standardizing the execution platform implies that the format of the code inside the packets is also agreed upon. But users and developers can code their own custom protocols in the packets. The code may describe a new protocol for transporting video packets or it may implement a custom routing algorithm for packets belonging to an application. The ability to introduce custom protocols breaks down barriers to innovation and enables developers to customize network resources to effectively meet their application's needs. Standardizing the execution environment enables new protocols to be designed, developed and deployed rapidly and effortlessly. The protocol designer develops the protocol for the standardized execution environment and injects it into the network nodes for immediate deployment. This eliminates the need for a standards committee to design the protocol, for hardware vendors to implement it in their network devices, and for service providers to deploy the new devices in their networks. Application designers can write custom code that performs custom computation on the packets belonging to the application. This customizes network resources to meet the immediate requirements of the application. Thus the programmable interface provided by active networking enables applications to interact with the network and adapt to underlying network characteristics. Note that active networks differ from efforts underway in programmable networks. Active networks carry executable within packets while programmable networks focus on a standard programming interface for network control.

2.2 ACTIVE NETWORK OPERATION

Custom code is injected in the network in one of two ways. One approach is to download the code to the nodes separately from the data packets. The downloaded code carries the computation for some or all of the data packets flowing through the node. The code is invoked when the intended data packets reach the node. This is known as the discrete approach (da Silva et al., 1988; Huber and Toutain, 1997). The other approach is the capsule or integrated approach (Wetherall et al., 1999; Kulkarni et al., 1998). In this approach, packets carry the computation as well as the data. As the packets flow through the nodes, the custom code is installed at some or all the nodes in the network. These packets are called capsules or SmartPackets. These are the basic unit of communication in an active network. Applications inject SmartPackets in the network and active nodes process the code inside the SmartPackets. A hybrid of the discrete and integrated approaches is PLAN (Hicks et al., 1999). The nodes of an active network, called active nodes, provide the platform for SmartPackets to execute the custom code. Active nodes participate in the customization of the network by the applications instead of passively switching or routing data packets through the network.

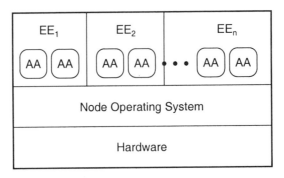

AA = Active Application   EE = Execution Environment

Figure 2.1 Basic Architecture of Active Node.

The active node, whose basic architecture is shown in Figure 2.1, provides the platform for execution that is called an execution environment (EE). An execution environment defines a virtual machine and a programming interface for users of the active network. Each EE provides a custom Application Programming Interface (API) for users to develop applications that run in the EE. An application developed to run in some EE on an active node is called an Active Application (AA). Active applications can utilize the EE's API to make informed decisions about its data or customize the node's behavior.

An active node can have multiple EEs. This enables the creation of a backbone of active nodes spanning different research institutions and yet enables researchers to independently investigate different execution models and features. The architecture of an active node allows support for and arbitration between multiple EEs through a common abstraction layer called a Node Operating System (NodeOS). The NodeOS provides the basic common functionality and access to resources shared by the EEs such as processing, memory and bandwidth. The NodeOS performs the tasks of resource management on behalf of the EEs and also protects them from the effects of the behavior of other EEs on the same active node.

To enable interoperability between active nodes that support different EEs, the active network community has defined a new protocol called the Active Network Encapsulation Protocol (ANEP) (Alexander et al., 1997). The ANEP header format is fixed and known to all EEs. The ANEP header contains an identifier field that uniquely identifies the EE that can process the current SmartPacket. Execution environments are registered with the Active Network Assigned Number Authority (ANANA) and they are assigned distinct numeric identifiers. SmartPackets from different execution environments can have different formats, and the identifier in the ANEP header enables active nodes to de-multiplex them to their correct processing environment. There are a number of EEs developed for active nodes to study various issues such as resource allocation, service composition, and security. The following section discusses features of an execution environment that are currently being explored. A few popular EEs are listed in Table 2.1.

Table 2.1. Table of Execution Environments.

EE Name	Institution	VM platform	Approach
Magician	University of Kansas	Java	Capsule
ANTS	Massachusetts Inst. of Technology	Java	Capsule
Switchware	University of Pennsylvania	Caml	Hybrid
Netscript	Columbia University	Netscript	Discrete
SmartPackets	BBN	Assembly code	Capsule
CANeS	Georgia Tech.	Unity	Capsule

Caml is a strongly typed functional programming language from the ML family of languages.

2.3 PROPERTIES OF EEs

As mentioned in the previous section, researchers have developed many different EEs that concentrate on particular facets of active networking mechanisms. This provides an opportunity to enumerate and classify properties that an execution environments possess to make code transport and execution possible. This section discusses the following principal features of EEs:

- Code download
- Implementation platform and language
- Composition
- Security and authentication
- Resource allocation and monitoring
- Inter-packet communication

An active network enables users to inject custom code into the network. There are different approaches for transporting code from the user to the active nodes. The earlier section alluded to two primary strategies: the integrated approach and the discrete approach. In the integrated approach, SmartPackets carry program code as well as data as they traverse through the network. In the out-of-band or discrete approach, program code is downloaded to the active nodes before sending the data.

The vision of active networks is that of a framework of networking components from which services are composed to provide users with a flexible, customizable networking infrastructure. Active networks will allow the composition of user-injected services with those that are pre-configured at the network nodes, enabling users to develop highly customized solutions in the network to handle their application data. The EE has to be able to provide modular service components for users to use as building blocks. Users will compose their own custom framework by using the "glue" provided by the EE in which their code will execute. The glue can be in the form of library modules, a scripting language or an API. The glue language itself can be a high-level language like

Java, C++ or C, or it can be a scripting language like Perl or Tcl. In this case, developers do not have to go through a learning curve, and this enables them to reuse existing code and integrate it into the active network framework. The glue language can also be custom designed for active networks. The advantage of a custom language is that the language can be designed by keeping in mind the unique requirements of active packets and the constraints that may need to be imposed on their functionality.

Components should have a formal, well-defined interface that abstracts the inner workings, yet presents the user with a rich representation of its behavior. This is defined as introspection. Introspection enables a user to access an implementation state and work with abstractions to it. This increases re-usability because the component can be reused in different protocol frameworks. The property of intercessation enables a user to modify the behavior of an abstraction. The user can refine existing behavior or adjust the implementation strategy to improve performance. This is usually achieved through the mechanism of inheritance in object modeling. This enables users to customize previously developed active network components, tailoring them to the needs of the application without having to write new components from scratch.

Execution environments can provide explicit security mechanisms for the execution, authentication and authorization of packets. Minimally, EEs have to execute packets within a sand-boxed environment that is protected from unintentional and malicious incursions by other packets. Additionally, EEs may implement trust mechanisms to authenticate various properties of the code contained in the packets, such as its owner-ship, creator, source, rights and privileges.

The primary resources of interest in an active network, as identified in the Active Network Architecture document discussed in Chapter 3, are transmission bandwidth, compute cycles, and storage. The EE may require mechanisms to monitor resource consumption by packets. It must not be possible for a malfunctioning or malicious piece of code to prevent other users from receiving adequate resources. An EE may also enforce "fairness" criteria for execution of the packet code.

An interesting feature of some EEs is that of SmallState, a named cache that enables SmartPackets to leave information at a node that can be retrieved later by other Smart-Packets. SmallState thus becomes an inter-packet communication (IPC) mechanism, providing a mechanism for packets to talk to each other and implement collaborative behavior in the network. The extent of the IPC depends on the control mechanisms defined by the EEs for their SmallState. In some EEs, such as Magician, SmartPackets explicitly create, modify and destroy SmallState. Access policies are defined for SmallStates at the time of creation. This includes defining access control lists or access masks for accessing, reading and writing SmallState.

2.3.1 ANTS: Active Network Toolkit

In the Active Network Toolkit active messages are called capsules. Capsules are identified by a unique type based on the MD5 message digest of the capsule code. When a capsule arrives at a node, a cache of protocol code is checked for the presence of the capsule code which may already reside at the current node. If the code does not exist on the current node, a load request is made to the node from which the capsule originated. The originating node receives the load request and responds with the entire code. The current node then incorporates the code into its cache. Processing functions may update capsule fields and store application-defined information in the shared node cache of soft-state. ANTS (Wetherall et al., 1999) provides a mechanism for protocol composition using the *CodeGroup* attribute of the ANTS capsule. Code belonging to the same protocol suite (or stack) shares the same *CodeGroup* attribute. Protocol modules are demand-loaded in separately into the active node prior to capsule execution. A wrapper module (which is the main class for the protocol) puts together the protocol modules in the manner desired by the programmer. ANTS is written in Java; therefore it supports modularity, inheritance and encapsulation. The base implementation available from MIT does not support any additional security features beyond those provided by the Java virtual machine. Other groups, such as the Trusted Information Systems Laboratory at Network Associates, Inc., have modified the ANTS prototype to add security and authentication mechanisms (Murphy, 1998).

2.3.2 Magician

Magician (Kulkarni et al., 1998) is a toolkit that provides a framework for creating SmartPackets as well as an environment for executing the SmartPackets. Magician is implemented in Java[1] version 1.1. Version 1.1 was primarily chosen because it was the first version to support serialization. In Magician, the executing entity is a Java object whose state is preserved as it traverses the active network. Serialization preserves the state of an object so that it can be transported or saved, and re-created at a later time. Magician utilizes a model in which an active node is represented as a class hierarchy. Every protocol is derived from an abstract base protocol. Every active node provides some basic functionality in the form of certain default protocols, e.g., routing. Users may prefer to utilize these default protocols if they do not wish to implement their own schemes. To foster privacy and safety, a unique state is created for each invocation of the protocol. A state that is common to all invocations of a protocol is inviolable and is accessible only to users that have appropriate authorization. Providing each user with a protected copy of the state enables the user to customize his own state if necessary. Secondly, users are permitted to install their own protocols. A user creates a new protocol from scratch by extending the base abstract protocol. When the protocol is carried over to the active node by a SmartPacket, a new branch is created for that protocol. Alternatively, the user extends an existing protocol class at the active node and customizes it to implement application-specific behavior as shown by the example of a user-defined routing algorithm in the figure. Thus, he is able to inherit certain basic properties of the parent protocol and provide minor or major behavioral modifications to tailor it to his application's requirements. In this model, the active node is thus perceived as an object with behavior that can be modified dynamically. SmartPackets are treated as objects that carry custom protocols in the form of executable code, which when installed at the active nodes, modifies their behavior to suit application requirements.

The class hierarchy model for active nodes is advantageous in many respects. It provides the user with the ability to customize default functionality at the active nodes using the extension-by-inheritance philosophy. The inheritance model also provides default functionality for user SmartPackets that choose not to implement their own. That enables the active node to emulate behavior of traditional nodes to support transport of conventional packets. The unique state space provided to each invocation of a protocol corresponds to the instantiation of an object from its class definition. This affords privacy and safety to the SmartPacket executing in that state. It also provides an isolated, secure environment for the execution of user code so that the active node can monitor and regulate its activities. The class-model approach differs from the data flow composition model, wherein protocol functionality is activated only by data flow. In the class model, instead of data being transferred between protocol components, execution control is transferred between the protocol components. Protocol components have access to the user data and can choose to act on the data or perform other activities. The class model also exhibits three of the five properties listed earlier: modularity, introspection, and intercession. Modularity is achieved by breaking down communication functionality into distinct protocol components. The functionality of the components is defined as a Java class. The property of introspection is satisfied when protocol components are designed as suitable abstractions and specified with well-defined interfaces that promote re-use. Using the principles of inheritance and aggregation, these components are brought together to form a composite that implements the functionality desired of a service to be injected in the network. This satisfies the intercession property.

2.3.3 NetScript

NetScript (daSilva et al., 1998) is a dataflow programming language in which computation consists of a collection of interconnected programs that process data streams flowing through them. The language supports the construction of packet processing programs, interconnecting these programs and allocating resources to them. NetScript programs communicate with each other through their input and output ports. The arrival of data at an input port triggers execution of sequential event handling code inside the program. NetScript is not a full-blown language and lacks traditional programming constructs like iteration control, functions and data types. Therefore, the programs (called boxes in NetScript) themselves are written in languages such as C and Java.

NetScript provides constructs for gluing together the protocol processing boxes and thus is viewed more as a coordination language. Protocol composition is achieved by using the NetScript constructs to define and establish interconnection between the various boxes. Dynamic composition is achieved using the concept of box templates. A box template describes the input/output port signature and the internal topological structure of the dataflow program. The port signatures define the types of messages that the box accepts and generates. The topological structure describes the interconnections between the internal boxes. It is possible to create one or more instances of this box template on active engines across the network. A new protocol framework is created by defining a composite box template that describes the protocol components and their interconnections. The protocol components themselves can be box templates or they can be primitive programs. The framework thus constructed is deployed in the network by dispatching it to individual NetScript engines at the network nodes. When a NetScript engine receives a new box template, it creates a running instance of it by examining its port specification and internal topological structure. The engine then loads and instantiates each internal box, makes appropriate connections between box ports, and starts up the new instance to allow protocol messages to flow through it. The composition mechanism is rich enough to enable composition of layered protocols or the extension of protocol components with new features. However, the extension mechanism is not as rich as the extension mechanism of an inheritance model. It is not possible to override only a few properties of a box. The box must be used with its established properties or a new box must be defined with the unwanted properties replaced by the desired ones.

2.3.4 PLAN

Switchware (Alexander et al., 1998) provides protocol composition using a two-level architecture that consists of switchlets and user programs in active packets. Each switchlet provides services in the form of methods manipulating protocols implemented for that switchlet. The switchlets services are implemented in a high-level language and are dynamically loaded at the network nodes. Active packets are user packets that invoke the services provided by the switchlets. The code in the active packets is written in a language called PLAN (Hicks et al., 1999) specifically designed for active networks. PLAN is a stateless, strongly typed, functional-style lambda calculus system with pointer-safe access, which addresses safety and security concerns. PLAN programs also make statements about resource needs and bounds. A PLAN program is initially assigned resource units which are depleted as it executes, or it spawns new PLAN programs. Since resource units cannot increase, they eventually become exhausted, which guarantees program termination. PLAN packets contain fields to help manage the packet: evaluation destination, resource bound, and routing function. While PLAN is implemented in Java, the programming language PLAN provides to the user is based on ML with some added primitives to express remote evaluation. PLAN programs can allow resident data to remain on a node after the program terminates.

2.3.5 CANEs

The Composable Active Network Elements (CANEs) execution environment was developed after the Architectural Framework Group and the NodeOS Working Group put together their specifications. Thus the CANEs environment is a NodeOS specification-compliant EE that sits on top of the Bowman node operating system. CANEs uses a slot processing model that comprises two parts: a fixed part, also called the underlying program, that represents the common processing applied by an active node on every packet passing through the node, and an injected program implementing user-specified functionality. The injected program is bound to a specific *slot*, which, when raised, causes the bound program to be executed. The underlying program provides a basic service like forwarding, that can be chosen from amongst that are well-known and pre-installed on every active node. A set of injected programs is then selected to customize the underlying program. These injected programs can be available at the active node, or they can be downloaded from a remote site. CANEs thus uses a discrete approach for downloading code. Since CANEs is built on top of Bowman, which uses the NodeOS specification, resource allocation and security issues are handled by Bowman. Bowman also implements the IPC mechanism, called state-store, to exchange data between packets without sharing variables. CANEs programs are written in custom language called CANEs User Interface (CUI) language. It comprises distinct computation and communication parts. The computation part is a directed acyclic graph (DAG) in which the underlying program sits at the root of the graph and the injected programs are at the leaves. The arcs in the DAG identify the slot to which the child is bound. Multiple programs can be bound to the same slot. Binding of the injected programs to the slots is done when the computation part is parsed by CANEs. The communication part of the CUI program identifies a routing schema for the user. The user may also specify a set of classifier rules that are used to select incoming packets that the flow acts on.

2.4 RESEARCH DIRECTIONS

This new paradigm in networking has generated excitement, and research efforts in this area are growing rapidly. Academic institutions working on active networking include Massachusetts Institute of Technology, University of Pennsylvania, Columbia University, University of Arizona, University of Kansas, University of California, Berkeley, among others. GTE/BBN, General Electric, Network Associates, Boeing and SRI are among active industry participants in this research. A list of participants currently sponsored by DARPA can be obtained from their web site.[2] This section describes the various initiatives taken by the community to promote cooperation and facilitate research in this new area. In order to understand the picture of the research landscape, the various projects and research efforts are classified into different categories. This should enable the reader to get a better picture of the work performed on the topic of his or her choice.

2.4.1 Working Groups

The DARPA Active Networking Community has formed several working groups to study various facets of this new model. Currently, the following groups are in existence:
- Architectural Framework Working Group (Calvert, 1998)
- Node Operating Systems Working Group (Peterson, 1998)
- Working Group on Security (Murphy, 1998)
- Composable Services Working Group (Zegura, 1998)

The *Architectural Framework Working Group* focuses on describing a general reference model into which active networking architectures can be mapped. The objectives of this working group are to lay down fundamental tenets and objectives and to define major components and interfaces that make up an active node and discuss architectural features common to network services built using active networks.

The aim of the *Node Operating Systems (NodeOS) Working Group* is to define an interface between the underlying operating system at each active node and its execution environments. The goal is to identify a minimal fixed point for the interface to enable developers of execution environments to present a range of abstractions about the underlying resources to the SmartPackets.

The emphasis of the *Composable Services Working Group* is on the interaction of services within an active network. The objectives of this working group, as listed in their draft document, are:

- to enumerate features that will be present, in some form, in any composition method
- to provide examples of composition methods based on common features
- to outline common issues that go beyond basic functionality and are amenable to solutions that require sharing

The *Security Working Group* is working on establishing mechanisms that enforce reasonable access and control mechanisms within components of the active network as well as recommended primitives available to active applications for interacting with the security architecture. Their goal is to define a guide for developing secure active network applications and creating a reference model for the provision of security features in the active network. While at the time this book was written, there was no formal working group on network management, there has been some work by interested parties in putting together a working document on a new network that leverages the active nature of packets within active networks to manage the active network. While the focus of such a group would be upon managing active networks, one can imagine the extremely flexible characteristics of active networks used as an overlay network to manage a legacy network.

2.4.2 Testbeds

Efforts are underway in the DARPA active networking community to create a large-scale worldwide testbed network, called the ABONE. Currently, the ABONE has a growing number of nodes spread across the continental United States. The target is to make the ABONE scalable to 1000 nodes spanning the world. Active nodes in the ABONE exchange SmartPackets encapsulated inside UDP datagrams. The on-the-wire format of a SmartPacket is defined in the Active Network Encapsulation Protocol (ANEP) document (Alexander et al., 1997). As explained in an earlier section, the ANEP protocol enables an active node to run multiple execution environments and yet demultiplex incoming SmartPackets to their correct processing environment.

The ABone currently has two primary types of users, Execution Environment developers and Active Application developers. The ABone provides a large scale active environment for Execution Environment developers to implement and test their EEs as well as develop and run Active Applications. EE developers have stricter security requirements because EEs have signicant control of the NodeOS, while AA developers can be given relatively free access to test active applications on any supporting EE.

The ABone provides a sharable resource in which AAs and EEs can be tested concurrently on a large scale, while attempting to maintain high availability in such an experimental environment. One mechanism supporting high availability is permanent EEs. These are EEs which will always exist on a node and which AAs can depend upon. Temporary EEs can reside on a node in an on-demand basis.

The base of the ABone is assembled from existing networks, partly from the existing DARPA CAIRN testbed and partly from AN research sites. The connection between indirectly connected active nodes occurs through Internet overlays. The ABone nodes contain many operating systems -- both general purpose and active Node Operating Systems to support the various EEs. The ABone provides an ideal place to experiment with configuration and management capabilities for active networks on a larger scale -- on the order of hundreds to thousands of active nodes. The ABone is divided into Core nodes and Edge nodes. Core nodes exist as a public resource and must be always available and reasonably stable. Edge nodes are directly under the administration of individual AA or EE developers. As such, Edge nodes, are not required to have the same stability or availability as Core nodes. Access to Core nodes occurs through Edge nodes.

The ABone requires the use of the Active Network management daemon, Anetd, to accomplish decentralized management of EEs. Anetd was designed to add new functionality for managing EEs while requiring minimal change to the EE itself. Anetd runs as a user process and has minimal installation requirements. Anetd also demultiplexes active network packets to the correct EE when multiple EEs reside on a node. This demulitplexing functionality will likely be migrated into the NodeOS in the future.

Because of the interaction required of the base ABone sites at CAIRN and separately administered active research sites, the ABone Coordination Center or ABOCC has been established to coordinate administration. The ABOCC is a DARPA-funded activity at ISI, and SRI and provides registration, maintenance of master files used with Anetd, status monitoring, trouble-shooting and diagnosis, and an Active Application repository.

Security is addressed in several levels: NodeOS, EE, and AA. At the NodeOS level, unix mechanisms such as accounts, file permissions, and process permissions are utilized. Anetd has its own security for deploying and managing EEs. The NodeOS should prevent the EE from disrupting other EEs, but must assume the EE will try to subvert the NodeOS itself. At the EE level of security, the EE must protect itself from AAs. At the AA level, the EE should attempt to keep the AA from subverting the EE and NodeOS; however, the details of the degree to which this is enforced is EE specific. An EE developer is identifed by a principleID and authenticated by a public/private key pair. An access control list is used to enforce who may install or modify an EE. Additionally, a trusted code server contains a list of code servers from which EE code can be retrieved by Anetd.

2.4.3 Research Areas

Figure 2.2 shows some of the areas in which active networking is likely to have a major impact. Ongoing research efforts target a wide range of issues pertinent to active networking, including those shown in the figure.

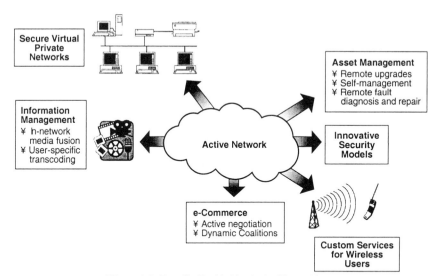

Figure 2.2. Benefits Enabled by Active Networking.

Broadly, research projects can be classified into the following areas:
- Network Management
- Security
- Integration of Mobile and Wireless networks
- Enhanced Network Services
- Quality of Service
- Mobile code and safe languages

Table 2.2 provides a snapshot of some of the prominent research efforts in each of the areas outlined above. This list is by no means exhaustive. There are other internal efforts at other institutions that have not received much attention. The rest of this section provides a high-level description of the projects listed in the table below.

2.4.3.1 Network Management

Currently, MIT, GE CRD, Columbia and University of Kansas are contributors to research in this area. The goals of the "Robust Multi-Scalable Networks" project (Legedze et al., 1998) at MIT are:
- to develop solutions that have the ability to monitor, manage and optimize the network based only on high-level external guidance; and
- to provide the ability to scale without compromising administrative and security functions

Table 2.2. Table of Research Projects

Research Area	Project	Organization
Network Management	· Robust Multi-Scalable Networks · NESTOR - Self Configuring Active Networks · Active Virtual Network Management Protocol	· MIT · Columbia Univ. · General Electric CRD
Security	· Secure Active Network Prototype · New Cryptographic Mechanisms · Dynamic Interoperable Security Mechanisms · Active Intrusion Detection and Response	· Network Associates (TIS) · Network Associates (TIS) · Univ. of Illinois · NAI Labs and Boeing
Mobile and Wireless	· SpectrumWare · MAGIC-II: Adaptive Applications for Wireless Networks · Proactive Infrastructure	· MIT · Univ. of Kansas · Univ. of California, Berkeley
Enhanced Network Services	· Active Web Caching · Composable Active Network Elements · Active Network Applications · Specialized Active Network Technologies for Distributed Simulation (SANDS)	· UCLA · Georgia Tech. and Univ. of Kentucky · Univ. of Washington (TASC) and Univ. of Massachusetts · Univ. of Washington (TASC) and Univ. of Massachusetts
Quality of Service	· Critical Technologies for Universal Information Access · Active Quality of Service	· MIT · General Electric (CRD)
Mobile Code and Safe Languages	· PLAN · NetScript · Liquid Software	· Univ. of Pennsylvania · Columbia Univ. · Univ. of Arizona

The NESTOR project (Yemini et al., 2000) at Columbia aims to create self-managed and self-organizing networks that enable vendors, system and network administrators to script configuration changes and consistency rules for network resources using a configuration policy langauge (CPL). The idea is to organize all network resource information in a universal resource directory that supports recoverable configuration change transactions and enforces respective configuration policies and enables automatic incorporation of new resource types and policies as the network changes. The goal of the AVNMP project (Bush, 1999; Bush, 2000) at GE's Corporate Research and Development (CRD) is to develop networks capable of predicting their own behavior and then using this capability to develop predictive network management techniques. The MAGIC-II project (Kulkarni et al., 1998) at the University of Kansas has developed novel algorithms for ensuring of active networking services injected into the network. They have leveraged active networking to enhance traditional distributed survivability mechanisms to create frameworks for user-injected services.

2.4.3.2 Security

The dynamic and proactive nature of an active network increases the security risks of unauthorized or destructive modification of the overall network behavior. Security issues are being considered and secure components developed now rather than being retrofitted after active network designs have solidified. Network Associates (TIS Labs) is investigating, as part of their project Secure Active Network Prototype Project (Murphy, 1988), security issues applicable in an active network, defining security requirements, and developing mechanisms to meet the requirements. Network Associates Inc., in collaboration with Boeing, is also studying development of new and response capabilities such as vulnerability scanning, attack tracing, and repair and recovery using active network technology. Cryptographic techniques in use in current networks follow the existing Internet paradigm of homogeneous end-to-end protection and are inadequate to the task of providing security for the full power of active network features. Another group at TIS Labs is investigating[3] expanded cryptographic requirements of active networks. The goal is to identify the constraints on cryptographic techniques particular to active networks and to develop and implement new or extended cryptographic techniques needed to meet the unique requirements and constraints of active networks.

2.4.3.3 Integration of Mobile and Wireless Elements

One of the major challenges of the current Internet is to be able to seamlessly integrate wireless and mobile elements in the network and provide enhanced services for wireless/mobile applications. Researchers at the University of California, Berkeley, the University of Kansas and MIT are among groups that are focusing on leveraging active networking to meet the above goals. The Proactive Infrastructure (Ninja) project[4] seeks to provide mechanisms for graceful degradation during times of intermittent connectivity and disconnected operation. The goal is to provide automatic on-demand data transformation to enable mobile and wireless users to access all services, by providing format conversion in the connection path. The Spectrum Ware project[5] at MIT aims to design solutions that enable wireless devices to adapt quickly to specified changes in functionality. Their approach is to move the hardware-software and digital-analog boundaries closer to the antenna, and perform all functions in software to handle unanticipated conditions.

2.4.3.4 Enhanced Network Services

The Active Web Caching project[6] at UCLA looks at an innovative model for caching web content in the network. Using the programmability of an active network, researchers are trying to create autonomous self-organizing web caches and servers distributed across the network. Client requests are routed to the appropriate servers based on their content. Researchers at the University of Washington have undertaken to develop three challenging large-scale distributed applications that benefit from the active network infrastructure. The first is a new wide area naming system, called Active Names, that binds names to client-specific and service-specific programs to provide flexible resource location and transformation. The second is a distributed infrastructure for measuring the performance characteristics of (inactive) Internet routers and links using coordinated measurements from active routers. The third application, called Detour, provides multi-path adaptive routing and aggregated congestion control to provide an order of magnitude improvement in end-to-end latency, bandwidth, and predictability relative to the passive Internet.

2.4.3.5 Mobile Code and Safe Languages

Research in this area focuses on the development of mobile, safe and secure languages for active networks. Active network languages provide the programming abstraction for users to access network state, customize network resources and implement custom services in the network. The University of Pennsylvania is working on a language called PLAN, which stands for Packet Language for Active Networks. PLAN is a language for programs that form the packets of an active network. PLAN is a type-safe, lightweight and mobile language with restricted functionality. PLAN behaves like a scripting language to compose or "glue" together trusted, well-defined service components resident at the active nodes. The design of PLAN enables it to make statements about the properties of programs written in the language. It is possible to guarantee properties such as the bounded consumption of network resources and the eventual termination of PLAN programs. NetScript (daSilva et al., 1998), developed at Columbia University, is a language system to construct, dynamically deploy, and execute active-element programs in network nodes. NetScript provides mechanisms to construct active elements to process packet streams. It supports end-to-end composition by interconnecting active elements in multiple nodes to form Virtual Active Networks (VANs). These active elements allocate node and bandwidth resources to enable end-to-end processing, reflect end-to-end priorities and QoS. VAN supports virtual network topology and provides mechanisms to implement security and protection for services deployed in a VAN.

2.4.3.6 Quality of Service

The objectives of the Universal Information Access project (Massachusetts Institute of Technology, 1999) at MIT are to identify fundamental architectural building blocks critical to a ubiquitous, universal network. Their objectives include the development of a scalable technology for network-level quality of service control using a new approach in allocating network resources to classes of users, including naming and name resolution. The project at General Electric (CRD) studied an active-network based mechanism for MPEG video. The intent was to evaluate the performance of an active control mechanism that allows individual packets to determine their own delay, on a hop-by-hop basis, in order to minimize jitter. This means that individual packets can vary their jitter control algorithm based on their overall value to the final product. This not only provides a finer resolution of control but also enables efficient utilization of network resources.

2.5 SUMMARY

Active networking offers a new paradigm for networks in which the packets can contain code fragments in addition to data that is executed on the intermediate nodes (switches and routers) of the network. The code carried by the packet may extend or modify node behavior enabling users and applications to tailor network resources for their immediate requirements. The goal of active networking is to increase the flexibility and customizability of the network and to accelerate the pace at which new networking service can be developed and deployed. Active networks will have an impact on almost all aspects of network operation and control. The ability to develop and deploy new, customized services benefits ad-hoc and wireless networks and mobile devices. Network infrastructure costs are reduced through self-configuration and management. The programmability of the network enables the deployment of innovative fault-tolerant algorithms to ensure survivability of network services and an improvement in the reliability of the network. Security is integrated "from the ground up" to create secure VPNs. Active networking also provides specialized processing for handling QoS requirements when used in conjunction with other networking technologies like ATM and SONet that are built for fast-path transport of data.

2.6 EXERCISES

1. What is ANEP and why was ANEP defined?
2. What are the discrete, integrated, and capsule approach?
3. What is Smallstate? When is it used?
4. In the Magician EE, how are protocols composed?
5. Describe the model used by the Netscript Execution Environment to compose protocols.
6. Describe the PLAN and CANEs composition models.
7. How is PLAN able to guarantee properties such as bounded consumption of network resources?

Notes:

[1] James Gosling and Henry McGilton, The Java(tm) Language Environment: A White Paper, Sun Microsystems.

[2] http://www.darpa.mil/ito/research/anets/projects.html.

[3] S. Murphy, D. Balenson and T. Samples. New Cryptographic Techniques for Active Networks,
http://www.nai.com/nai_labs/asp_set/network_security/an_sanp.asp.

[4] E. Brewer, R. Katz, D. Culler, A. Joseph, A. Quinn and T. Lessard-Smith. Proactive Infrastructure, http://ninja.cs.berkeley.edu.

[5] John Guttag, Thomas Low and Paul Powell. SpectrumWare,
http://www.sds.lcs.mit.edu/SpectrumWare/.

[6] Lixia Zhang. Adaptive Web Caching, http://irl.cs.ucla.edu/awc.html. David Clark and Karen Sollins. Universal Information Access, http://ana-www.lcs.mit.edu/darpa/darpa-1998.html, Project Summary, 1998.

II

ACTIVE NETWORK ARCHITECTURE

3

ARCHITECTURAL FRAMEWORK

There is a general consensus that network architecture must be made programmable to enable application-specific customization of network resources and the rapid introduction of new services in the network. As a result, there are two schools of thought on the approach to be taken to make networks programmable. One effort is spearheaded by the OpenSig community, which argues that networks can be architected using a set of open programmable interfaces, opening access to network devices like the switches, routers and hubs, thereby enabling third-party developers to devise custom solutions for their applications. The other approach is that of active networks in which packets can contain custom code that executes in the execution environment provided by the nodes of the network. Thus an active network is no longer node-centric but packet-centric; that is, the code in the active packets is executed at the intermediate network devices as the packet travels through the network. The result is that customized processing is accomplished as the packet is transmitted through the network. This enables experimentation and the rapid deployment of new services.

After the concept of active networking was initially conceived, researchers at the Massachusetts Institute of Technology developed a small proof-of-concept prototype using a Tcl shell to demonstrate the potential of the new idea. This sparked interest among researchers initially in the United States (primarily because of the DARPA effort) and later world-wide. This is evidenced from the papers submitted to the March 2000 issue of the IEEE Communications Magazine and the participation in the International Conference on Active Networks (IWAN). There were a considerable number of research efforts that attempted to develop complex and varied execution platforms to study different facets of the model. This led to a plethora of custom execution platforms on small-scale active networks that were limited in their deployment to their own laboratories or a few peer institutions. However, the prototypes were limited to testbeds in individual research laboratories, which did not bode well for the deployment of active networks on a global scale. Therefore, in 1998, researchers across the active networking community challenged themselves to develop plans to deploy active networks on a large scale. The principal hurdle towards this goal was the total lack of interoperability between the various execution platforms. The community decided that while the richness in diversity is desirable, it is necessary to develop a scheme that enables diverse active network prototypes to communicate with each other. The solution was to define a skeleton framework for an active node that divided its functionality into layers and to define the scope and properties of each layer. This enabled researchers to collaborate in the design of an active network in which active nodes can host several different execution platforms and yet can be connected to each other to provide an active network backbone. The framework is defined in a working document called the "Architectural Framework for Active Networks" (Calvert, 1998). This chapter discusses salient issues described in the document. It adopts a didactic and dialectic form for discussion to enable the reader to gather a better perspective in understanding the issues behind the decisions taken. The discussion here is neither meant to be exhaustive nor is it complete because the document itself is a work in progress. The serious reader is encouraged to go through the document. The intent is to provide the springboard for the reader to understand the issues described therein.

Imagine a time in the future where someone digs up a crusty old technical document. This document makes its way into the hands of a few bright minds of the time who instantly recognize it to be the foundation work of the late 20th century on a fledgling technology called "active networking" that evolved into the current communications infrastructure. These bright minds parse the document and attempt to figure out the reasoning behind the decisions outlined in the manuscript. The names of the characters in the dialog are purposely reminiscent of ancient Greece, foreshadowing the issues of rollback and tangled hierarchies to be discussed in later chapters.

3.1 THE DISCUSSION

The quotation that follows is about the power of the dialectic method in its pursuit of the study of truth through discussions based upon questions and answers. The remainder of this chapter attempts to use the dialectic format.

"The very things which they mould and draw, which have shadows and images of themselves in water, these things they treat in their turn as only images, but what they really seek is to get sight of those realities which can be seen only by the mind. "True" he said. "This then is the class that I described as intelligible, it is true, but with the reservation first that the soul is compelled to employ assumptions in the investigation of it, not proceeding to a first principle because of its inability to extricate itself from and rise above its assumptions, and second, that it uses as images or likenesses the very objects that are themselves copied and adumbrated by the class below them, and that in comparison with these latter are esteemed as clear and held in honor. "I understand" said he, "that you are speaking of what falls under geometry and the kindred arts." "Understand then," said I, "that by the other section of the intelligible I mean that which the reason itself lays hold of by the power of dialectics, treating its assumptions not as absolute beginnings but literally as hypotheses, underpinnings, footings, and springboards so to speak, to enable it to rise to that which requires no assumption and is the starting-point of all, and after attaining to that again taking hold of the first dependencies from it, so to proceed downward to the conclusion, making no use whatever of any object of sense but only of pure ideas moving on through ideas to ideas and ending with ideas." "I understand" he said, "not fully, for it is no slight task that you appear to have in mind, but I do understand that you mean to distinguish the aspect of reality and the intelligible, which is contemplated by the power of dialectic, as something truer and more exact than the object of the so-called arts and sciences whose assumptions are arbitrary start-ing-points. And though it is true that those who contemplate them are compelled to use their understanding and not their senses, yet because they do not go back to the beginning in the study of them but start from assumptions you do not think they possess true intelligence about them although the things themselves are intelligible when apprehended in conjunction with a first principle. And I think you call the mental habit of geometers and their like mind or understand-ing and not reason because you regard understanding as something intermediate between opinion and reason."

Plato, *The Republic*, 511a

Each section of the discussion that follows is introduced with an indented excerpt from (Calvert, 1998). We gratefully acknowledge and appreciate the fine work the DARPA Active Networks Working Group has done in producing this document.

3.1.1 Assertions

Proposition 1 *[Fundamental Tenets]*
The following assertions are considered to be "given":
- The primary function of the active network is communication and not computation. The network contains some nodes whose sole purpose is to allow sharing of transmission resources. Computation may occur, and in-deed computation services could be built upon the active network plat-form, but the platform itself is not designed to be a general-purpose system.
- The unit of the network is the packet (and not, say, circuits).
- Active nodes are interconnected by a variety of technologies, and this vari-ety will evolve continually. Therefore assumptions about underlying serv-ices must be minimized.
- Each active node is controlled by an administration, and no administration controls all active nodes.
- Trust relationships may vary between administrations. In general, trust is explicitly managed.

Socrates[1]: "Did you notice that there is an interesting assumption made by this document about active networking in general? For example, it explicitly states that the function of an active network is communication and not computation. It further adds that the platform is not designed to be a general-purpose distributed computing system. But we know that active network platforms are general-purpose execution environments. Does not the reality contradict the assumption?"

Glaucon: "I did not read it that way. The nodes present an execution environment for limited albeit general computation. That is not to be confused with a distributed comput-ing system. I would say the difference between a distributed computing system and an active network is that the former utilizes the latter, that is, the underlying network serv-ices, to run applications, whereas active networks manipulate the network services to provide application-specific services."

Socrates: "So you are saying that the distributed computing system utilizes underlying communication services to implement an application whereas an active network provides a platform for introducing new services on behalf of the application."

Glaucon: "Exactly. The vision is that the primary function of an active network is communication. Applications use the computation feature to customize the network."

Socrates: "But what if the distributed computing system used middleware services?"

Glaucon: "The limitation of middleware services was that they could either be loaded dynamically at the end-systems of a communication channel or loaded *a priori* on all or some nodes of the network. New middleware services could not be introduced in a ad hoc manner in distributed computing systems."

Socrates: "Then how is an active network different from ancient networks that concentrated on simple non-active packet delivery with minimal overhead? The ancient philosophy had been to keep the network simple and push all the computation to the end-systems. That had been considered a winning virtue for the Internet Protocol, much like 'motherhood' and 'apple-pie' in America."

Thrasymachus: "You must consider the fact that even the most ancient networks that claimed to be simple had to have some processing of packets. Packet serialization and deserialization, forwarding, error recovery, and congestion control all required some form of processing in even the simplest networks. One difference is that this processing was highly specialized and required many years for everybody to learn and agree to do everything the same way. This often contributed to a slow pace of innovation in the network infrastructure."

Socrates: "This is clear. But even if they progressed slowly, they should have at least all converged to the same system eventually. Did they not?"

Thrasymachus: "Unfortunately, one had to keep adding new revisions and protocols to compensate for the "simpleness" of their Internet Protocol. Each time a change was suggested, it was submitted as an Internet Draft and after a long period of time some of them would become Request for Comments (RFC). Each change or addition required a new Request for Comment. By the year 2000 there were nearly 3000 Request for Comments. No one person could keep track of them all. The network became too complicated to modify because it became difficult to keep track of all the inter-dependencies among the RFCs."

Socrates: "I remember that there were some attempts at that time to alleviate the complexity by introducing Programmable Networks. What was that and how did it differ from active networks?"

Thrasymachus: "Programmable networks viewed the network as a virtual machine with a common interface. This was to facilitate programming by making a common API for the network."

Socrates: "So again, you are saying that even in those times, it was recognized that networks must have some form of processing. Otherwise, why did they need to go to the trouble of standardizing an interface for programming the network? But then how does one decide how much processing to put into the network?"

Thrasymachus: "In the early days, emphasis was placed on throughput and response times for data passing through the network rather than on the usability of the data to the application and the flexibility to introduce new services in the network."

Glaucon to Thrasymachus: "You mentioned non-active data in ancient networks, yet this appears to be one of the oldest documents in which the possibility of including code within a packet was ever mentioned. It still maintains that the packet is the basic unit of communication. Circuits are specifically excluded. Any reason why?"

Thrasymachus: "Circuit switching involves setting up a fixed pathway before communication begins. This would greatly reduce the flexibility of the network. The goal is the opposite, an 'active' network. However, a circuit switched network could be emulated by an active network. Thus, no functionality is lost by assuming the packet as the basic unit of communication."

Socrates: "In fact this point leads to the next assumption that 'active nodes are interconnected by a variety of packet-forwarding technologies...'. By making minimal assumptions about the underlying link-level connectivity, active networks can be overlaid on the greatest number of networking technologies."

Thrasymachus: "I believe they were keeping their options open here. They did not know what type of packets they might end up creating."

Socrates: "Who controlled the active network? How secure was it?"

Thrasymachus: "It was assumed to be much like the ancient Internet: each node administered by an administrator, but no single administrator controls all nodes. Security is not discussed until closer to the end of this document, but apparently they realized that different groups would have varying levels of trust in one another and that such trust had to be enforced and managed explicitly."

3.1.2 Objectives

Proposition 2 [*Objectives*]

The active network provides a platform on which network services can be experimented with, developed, and deployed. That platform represents a "meta-agreement" among the parties concerned with the network (users, service providers, developers, researchers) regarding the fundamental capabilities built into the network, as well as the mechanisms through which they may be accessed and combined to implement new services. Potential benefits of such a capability include faster deployment of new services, the ability to customize services for different applications, and the ability to experiment with new network architectures and services on a running network. The general role of the architecture is to describe the form of the active network platform, its major components and interfaces. As usual, the intent is to be specific but to avoid ruling out possible solutions by over-constraining things. Thus, the architecture defines functional "slots" in the active network platform, which may be filled in various ways. This architecture has been developed with the following objectives in view:

- Minimize the amount of global agreement required, and support dynamic modification of aspects of the network that do not require global agreement.

- Support, or at least do not hinder, fast-path processing optimizations in nodes. In particular, it should be possible to construct an active node that can route plain old IP datagrams at speeds roughly comparable to those of "passive" IP routers.

- Support deployment of a base platform that permits on-the-fly experimentation. Backward compatibility, or at least the ability to fit existing network nodes into the architectural framework, is desirable.

- Scale to very large global active networks.

- Provide mechanisms to ensure the security and robustness of active nodes individually as well as of the active network as a whole. Robustness mechanisms should be independent of security, so that the consequences of the actions of even the most-authorized user are limited in scope.

- Support network management at all levels.

- Provide mechanisms to support different levels/qualities/classes of service.

Socrates: "The main and novel (for its time) objective was to minimize the amount of global agreement required within the network to implement and deploy a protocol or service."

Glaucon: "Yes, this objective appears to be the most salient. The remaining objectives, while important, had been objectives for most previous network architectures."

Socrates: "Supporting fast-path processing, backward compatibility, scalability, security, robustness, network management, and varying QoS were required of networks before active networks."

Thrasymachus: "But aren't there better ways to minimize the amount of global standardization than the approach proposed in this architecture? For example, don't programmable networks solve this issue?"

Glaucon: "Programmable networks help, but they only provide a common virtual machine or API for network programmers. Introducing a new protocol or service still requires that everyone agree on a standard and implement and deploy the standard everywhere it is to be used."

Thrasymachus: "But how does the active network architectural framework address the issue of rapid protocol implementation and deployment?"

Glaucon: "It appears further along in this document that as part of this framework, packets can carry and execute code inside the network. This would allow a packet to carry its own protocol. Thus an end-system could inject packets that behave in the manner intended. Specific protocol 'intelligence' resides in the active packets rather than on the individual nodes. The network nodes are general purpose processing devices with an API that the developer of active applications is aware of. The developer also has to be aware of a few other things. Suppose he or she wishes to reuse an existing active protocol or service. Why should the developer re-invent the wheel? In fact the active network developer would likely wish to compose a new service from a variety of existing active services."

Socrates: "Didn't this general purpose processing of packets we just discussed add tremendous overhead? Didn't this affect scalability? Didn't this open the network up to large security vulnerabilities? What about fairness? What if an active packet never releases the processor? How were active networks introduced without being entirely separated from the current networking technology of the day?"

Glaucon: "OK, slow down. These are all good questions that were mentioned in the objectives. Let's see how the document addresses these issues."

3.1.3 Active Network Architecture

Proposition 3 *[Architecture]*
All nodes in the active network provide a common base functionality, which is defined by the node architecture. The node architecture deals with how packets are processed and how local resources are managed. A network architecture deals with global matters like addressing, end-to-end allocation of resources, and so on. As such, it is tied to the interface presented by the network to the end user. The node architecture presented here is explicitly designed to allow for more than one such "network API." Several factors motivate this approach. First, multiple APIs already exist, and given the early state of our understanding it seems desirable to let them "compete" side-by-side. Second, it supports the goal of fast-path processing for those packets that want "plain old forwarding service." Third, this approach provides a built-in evolutionary path, not only for new and existing APIs, but also for backward-compatibility: IPv4 or IPv6 functionality can be regarded as simply another API.

The functionality of the active network node is divided between the (Execution Environments or just "environments") and the Node Operating System (NodeOS). The general organization of these components is shown in Figure 3.1. Each Execution Environment exports an API or virtual machine that users can program or control by directing packets to it. Thus, an Execution Environment acts like the "shell" program in a general-purpose computing system, providing an interface through which end-to-end network services can be accessed. The architecture allows for multiple Execution Environments to be present on a single active node. However, development and deployment of an Execution Environment is considered a nontrivial exercise, and it is expected that the number of different Execution Environments supported at any one time will be relatively small. The NodeOS provides the basic functions on which execution environments build the abstractions presented to the user. As such, it manages the resources of the active node and mediates the demand for those resources, including transmission, computing, and storage. The NodeOS thus isolates Execution Environments from the details of resource management and from the effects of the behavior of other Execution Environments. The Execution Environments in turn hide from the NodeOS most of the details of interaction with the end user. To provide adequate support for quality of service, the NodeOS allows for allocation of resources at a finer level of granularity than just Execution Environments. When an Execution Environment requests a service from the NodeOS, the request is accompanied by an identifier (and possibly a credential) for the principal in whose behalf the request is made. This principal may be the Execution Environment itself, or another party-a user-in whose behalf the Execution Environment is acting. The NodeOS presents this information to an enforcement engine, which verifies its authenticity and checks that the node's security policy database (see 3.1) authorizes the principal to receive the requested service or perform the requested operation. A companion document [Murphy, 1998] describes the active network security architecture; see also Chapter 6.

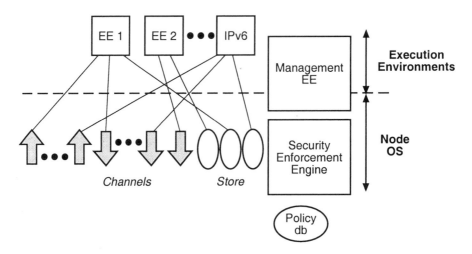

Figure 3.1. Active Node Components.

Glaucon: "In order to better understand the answers to your questions, Socrates, we need to begin with a better understanding of how the active packets interact with active nodes. We can use this active network framework document to trace the path of a hypothetical active packet."

Thrasymachus: "OK, let us assume the active packet is somehow injected into the active network and we are following it through the first active node in its path."

Glaucon: "Wait! We need to begin even further back than that. How was our hypothetical active packet created, what does it look like, and how did it choose to go to the first node within the network?"

Thrasymachus: "Good point."

Glaucon: "Let's say our hypothetical active packet needs to be injected into the network from an end-system. The exact structure of the packet is determined by the execution environment in which it is intended to execute. The target execution environment usually obtained the definition of the packet structure, created an instance of the packet, and forwarded it toward its destination."

Socrates: "Whoa. Does this mean multiple execution environments were built? Wouldn't this hinder attempts to create a standardized platform, which was one of the key objectives?"

Glaucon: "Yes, multiple execution environments were allowed initially, even on a single node. This was done purposefully to encourage research on various aspects of the then nascent technology. However, to promote collaboration between the various execution environments, a common overall format of the active packet and default functionality was agreed upon by the founders. The format was called the Active Network Encapsulation Protocol (ANEP) format. The ANEP header included a type identifier field. Each execution environment was assigned a specific Type ID. When a packet reached an active node, it was de-multiplexed to its correct processing environment based on the Type Identifier.

"To answer your second question, the ANEP header carried additional information that included forwarding, error handling, security identification, and fragmentation and re-assembly. The ANEP document stated that the packet could choose what would happen if such a scenario occurs. By default, the packet was forwarded towards the destination using the node's default routing policy. Error handling specified how the packet should be handled if an error occurs. The packet may be simply dropped, or an error message can be forwarded."

Socrates: "What if an older, non-active packet is received that did not use ANEP? I thought one of the objectives was backward compatibility."

Glaucon: "A legacy packet without an ANEP header was identified as such and routed to an execution environment that provided legacy protocol handling."

Socrates: "What about addressing? Did they have a common scheme specified?"

Glaucon: "No, addressing was not explicitly 'specified' by ANEP but it supported a place holder for defining one. Thus, new addressing schemes could be introduced into the network while still supporting old ones."

Thrasymachus: "That seems quite flexible and open; however, our packet must travel on some type of well-defined data link protocol at least. For how can a packet contain code such that it can verify its own integrity over a possibly noisy link?"

Glaucon: "You have identified a tough problem. In fact, the active network architecture document assumes that standard legacy protocols will provide a basis for active packet transport. Exactly how far down the stack active networking can be taken appears to have been an open issue."

Thrasymachus: "OK. We have our packet roughly defined with some guess as to how it will be transported, that is, anything from Ethernet to TCP. What else is inside our packet?"

Glaucon: "Typically, code that tailors the behavior of the nodes that the packet visits to the requirements of the application. The code may or may not have any constraints placed upon it. In some execution environments, the code was very restrictive in its capabilities while in some others, the only constraint was that it should be in a format supported by the execution environment. We are now ready for the next step in the life of our active packet. The packet determines its next hop to its destination and forwards itself appropriately."

Socrates: "That's nearly correct. However, a more sophisticated matching is required to determine what should be done with the packet. As we mentioned, the packet could be a legacy protocol packet and would need to be properly identified and sent to an execution environment that supported legacy protocols. Another possibility is that the packet may be simply 'cut-through' or sent straight to the output without any processing. The Type Identifier in the ANEP identifies the proper EE to execute the packet's code."

3.1.4 The Node Operating System

Proposition 4 *[NodeOs]*
To provide for quality of service, the NodeOS has scheduling mechanisms that control access to the computation and transmission resources of the node. These mechanisms isolate user traffic to some degree from the effects of other users' traffic, so that each appears to have its own virtual machine and/or virtual link. When channels are created, the requesting Execution Environment specifies the desired treatment by the scheduler(s). Such treatment may include:

- Reservation of a specific amount of bandwidth (transmission or processing), and isolation from other traffic.
- Isolation from other channels' traffic and guarantee of a "fair share" of available bandwidth, the available bandwidth is divided more or less equally among all channels in this class.
- Best effort, i.e. no isolation from other traffic in this class. Requests for channels with the first two types of service may be denied by the NodeOS when resources are in high demand. Note that input channels are scheduled only with respect to computation, while output channels are scheduled with respect to both computation and transmission."

Socrates: "If I understood the architecture correctly, the execution environments run on top of a generic node operating system or NodeOS. Why was the NodeOS required? Why couldn't the Execution Environments provide the same functionality?"

Thrasymachus: "This was part of the integration effort of the community at that time to build a large-scale network that encompassed all the existing Execution Environments or EEs. EEs are defined as lightweight virtual machines that provide the user interface. The NodeOS provides the functionality that is common to all the Execution Environments and also regulates the resources at the active node so that no one application or EE monopolizes a resource."

Socrates: "But why do the Execution Environments have to specify scheduling treatment for a channel from the NodeOS? Can the Execution Environments themselves not implement a scheduler that handles channels that it creates and monitors?"

Glaucon: "Interesting question. I think this mechanism was primarily created to avoid duplication of functionality since every Execution Environment would eventually need some kind of scheduler for handling packets in its channels. The NodeOS would be to provide efficient and fast processing at the expense of complexity."

Thrasymachus: "Notice that the active nodes also had to handle 'cut-through' flows. Such flows will definitely be processed much faster if the scheduling is performed in the NodeOS instead of its associated Execution Environment."

3.1.5 The Execution Environment

Proposition 5 *[Execution Environment]*
An execution environment defines a virtual machine and "programming interface" that is available to users of the active network. Users control that virtual machine by sending the appropriate coded instructions to the Execution Environment in packets. Execution of these instructions will in general cause the state of the Execution Environment and the active node to be updated, and may cause the Execution Environment to transmit one or more packets, either immediately or at some later time. The function of the virtual machine provided by any Execution Environment is not defined by the architecture. The NodeOS provides a basic set of capabilities that can be used by Execution Environments to implement a virtual machine. Thus, most aspects of the network architecture are defined by the Execution Environment.
Execution Environments allow users some control over the way their packets are processed. If an Execution Environment implements a universal virtual machine

– that is, one that can be programmed to simulate other machines – a user may provide a complete description of required packet processing. Other Execution Environments may supply a more restricted form of programming, in which the user supplies a set of parameters and/or simple policy specifications. In any case, the "program" may be carried in-band with the packet itself, or installed out-of-band. Out-of-band programming may happen in advance of packet transmission or on-demand, upon packet arrival; it may occur automatically (e.g. when instructions carried in a packet invoke a method not present at the node, but known to exist elsewhere) or under explicit user control.

Glaucon: "Following the path of our hypothetical active packet, assume that it is now handed off by the NodeOS to the Execution Environment for processing its contents. The question here is why do the Execution Environments not provide similar programming paradigms? Why are some Execution Environments more restrictive than others?"

Socrates: "I believe this is partly explained by the security considerations and partly by consideration of experimentation. A network administrator would be interested in controlling security levels of different nodes in his network. In those days, nodes that were deemed to be sensitive or critical had to be protected from malicious attacks and/or inadvertent mistakes in user code. To prevent this, the Execution Environment would allow only a restricted language set for the packets executing at the node. On the other hand, less sensitive or less critical nodes would have allowed the full programming interface."

Thrasymachus: "I see your point. But why do you think experimentation has anything to do with this?"

Socrates: "If you remember, the document says that a single active node can contain multiple Execution Environments. In the early days of experimentation, different groups studied various facets of the technology. Execution Environments that were developed to study security and resource allocation issues could have been more restrictive. Those groups that studied the impacts of the technology on applications to explore its potential would have developed Execution Environments that had a richer programming interface. As active networking matured, the two approaches converged into its final form that integrated security and resource monitoring capabilities with rich programming abstractions."

Socrates: "We have been discussing the path of our active packet as if it contained code; i.e., it was 'in-band'. What if the channel is 'out-of-band,' i.e., code is placed at the nodes before data is transmitted?"

Thrasymachus: "There is no difference as far as handling of the packet by the NodeOS is concerned. Both types of packets are handled the same way. The difference is in the Execution Environment's handling. Out-of-band packets do not carry execution code. Code download and execution are separated from the traversal of the packet through the network."

3.1.6 Fairness in Active Networks

Proposition 6 *[Principals, Channels and Scheduling]*
The primary abstraction for accounting and security purposes is the principal. All resource allocation and security decisions are made on a per-principal basis. In other words, a principal is admitted to an active node once it has authenti-

cated itself to the node, and it is allowed to request and use resources as de-scribed below. There are three primary resource abstractions: threads, and chan-nels. Threads and memory behave exactly as on a traditional OS, although a definition that is specific to active nodes may be defined at a future time. Chan-nels are defined in the next section. Principals create channels over which they can send, receive, and forward packets. Some channels are anchored in an exe-cution environment; principals use anchored channels to send packets between the execution environment and the underlying communication substrate. Other channels are cut-through, meaning that they forward packets through the active node –from an input device to an output device – without being intercepted and processed by an Execution Environment. Channels are in general full-duplex, although a given principal might only send or receive packets on a particular channel.

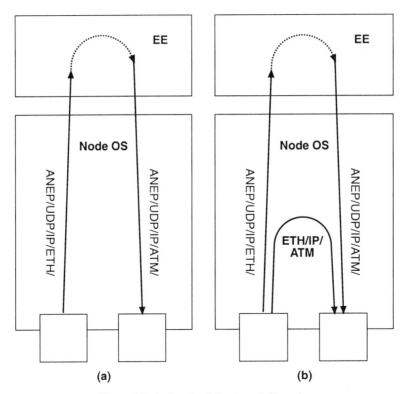

Figure 3.2. Anchored and Cut-through Channels.

It is helpful to define channels in terms of a more primitive building block-protocol modules – which implement some well-defined primitive communica-tion service. For example, ANEP, TCP, UDP, IP, an ethernet driver (ETH), and an ATM driver (ATM) are protocol modules with the obvious meaning. Protocol modules can be combined (stacked) to build more complex communication services. Note that a given NodeOS need not implement channels as stacks of protocol modules, although a NodeOS can leverage this implementation to support extensible channels. Figure 3.2 distinguishes between anchored and cut-through channels on an active node connected to a pair of network links (ATM and ETH). In (a), the principal has created a flow through the active node consisting of a pair of anchored channels (depicted as solid lines labeled /ANEP/UDP/IP/ETH and /ANEP/UDP/IP/ATM), and some computation that takes place in the context of the Execution Environment (depicted as a dotted line connecting the two channels). Note that the principal need not have created ANEP channels; it could have created channels to send and receive raw ethernet/ATM packets. In (b), the principal has created a flow through the active node consisting of the same channels as in (a), plus a cut-through channel that directly connects the ethernet link to the ATM link. This particular example includes IP behavior (e.g., decrementing TTL) on the cut-through channel. Packets flow over the cut-through channel without being intercepted and processed by the Execution Environment. In this example, the anchored channels (plus the in-Execution Environment processing) might by used by an active control protocol (e.g., an active version of RSVP) to exchange control messages with its peers. These control messages might then cause the cut-through channel to be created, with data messages forwarded over this data/cut-through channel. Principals create channels by specifying a set of attributes. These attributes include the modules that define the behavior of the channel, as well as other properties such as addresses, direction of packet flow, MTU, and quality of service requirements. In addition to specifying a set of channel attributes, the Execution Environment must also provide the NodeOS with enough information to allow the NodeOS to classify incoming packets and demultiplex them to the appropriate input or cut-through channel. Packet classification is not an issue for output-only channels. We envision this classification information being of one of two forms. Untrusted Execution Environments will specify a classification pattern that will then be interpreted by the NodeOS. Trusted Execution Environments may be allowed to contribute a code fragment that the NodeOS then runs to demultiplex packets to that channel. It is the NodeOS's responsibility to classify each incoming packet according to the information provided by the Execution Environments. Any packet that does not match an existing channel is dropped by the NodeOS; it is expected that some or all nodes will support a default channel implementing IP (or IP-like) forwarding semantics. The final responsibility of the NodeOS is to schedule channels for execution. Scheduling decisions are influenced by both the computational requirements of the channel – which depends on the protocol stack that implements the channel – as well as the bandwidth requirements corresponding to the channel's quality of service guarantee. For example, in Figure 3.2, two different channels have been created with IP processing for one output interface. One is the standard "best-effort" IP channel, while the other has an associated bandwidth reservation, and is only accessible to Execution Environment 2.

Thrasymachus: "We see here that our active packet is associated with a principal when it arrives at an active node. The principal governs a packet's interactions with the resources at the active node. Our packet is also part of a channel that is created by the principal to manage active packets."

Glaucon: "Wait a minute. What is the significance of a channel? If a principal governs the behavior of active packets for a type, why are channels required?"

Thrasymachus: "Principals describe the relationship of the originator of the active packets with the active node. They define what the active packets can or cannot do when they are at the active node. On the other hand, channels describe the properties of the active packets utilizing the resources of the node. So, one can say that principals describe the 'rights (what)' of the associated packets while channels describe the 'actions (how)'. Channels are described in terms of a composition of protocol modules. Each protocol module specifies a small piece of the 'how'. When these modules are put together, they define the properties of the channel. Of course, the principal defines if the channel can behave a particular way or not."

Socrates: "Then what about cut-through channels? How are they specified?"

Thrasymachus: "Cut-through channels are also specified the same way. The document distinguishes between anchored and cut-through channels. Anchored channels can define and use their own protocol modules. On the other hand, cut-through channels can only be defined in terms of protocol modules defined and exported by the NodeOS. Thus, anchored channels allow for greater customization than cut-through channels."

Glaucon: "Why is packet classification to channels required as part of the NodeOS functionality? Why not let Execution Environments accept packets from the NodeOS and route them to appropriate channels themselves?"

Thrasymachus: "I believe this policy was defined for two reasons. First, the designers wanted the Execution Environments to be lightweight and therefore attempted to move all common functionality into the NodeOS. Packet classification functionality is common to all packets irrespective of Execution Environment and hence is performed in the NodeOS. Second, this enables routing packets to cut-through channels. If the Execution Environments performed this task, then each packet would have to be sent up to the Execution Environment to be assigned to a channel. Support for cut-through channels would not have been possible."

Socrates: "So going back to our active packet, the NodeOS uses the classifiers provided by the associated Execution Environment to assign the packet to the correct channel. So, for example if the packet is a Magician Smartpacket, the NodeOS uses the packet type encoded in the Magician-specific ANEP header."

Thrasymachus: "That's right. Magician used non-active classifiers. Execution Environments could have provided filters to the NodeOS. The restriction would haven been that the code had to be written in the language supported by the NodeOS for the active node."

3.1.7 Active Network Interfaces

Proposition 7 *[Interfaces]*
The major interfaces of the architecture are those between the Execution Environments and the NodeOS, and between Execution Environments and applications. The former is of interest to node vendors and Execution Environment developers; the latter is of interest to application developers and Execution Environment developers. The interface between the Node OS and the Execution En-

vironments does not affect the bits on the wire, and is not directly visible to end users. Rather, it is local to the active node, and may be implemented in node-specific ways, like the system-call interface of an operating system. The specification of this interface consists of "system calls" (methods) provided by the Node OS to the environments. The NodeOS interface is currently under development, and will be defined in a companion document. The interface between the Execution Environment and the end-user application is by definition Execution Environment-specific. However, it must be precisely specified in order for an Execution Environment to be deployed. Some environments may offer services that are accessible via a generic, language-independent interface (a la sockets); others may require that application and network code be integrated, or even written in a particular language (e.g., Java). The intent is that this interface should be more or less independent of the underlying active network infrastructure. Each active node should support a special environment providing a maintenance interface for the Node OS and the common abstractions it supports. Such an interface would provide control over node policies and parameters, as well as maintenance of the common abstractions that are supported by the Node OS and made available to every environment.

Glaucon: "Some of the interface definitions confuse me. For example, consider the Execution Environment-NodeOS interface. The document states that this interface 'is local to the active node and may be implemented in node-specific ways'. Does this not mean that early Execution Environment developers would have to define multiple interfaces for their Execution Environment to enable it to run on different nodes?"

Thrasymachus: "Yes, that's true. However, in the early days there was work performed at the University of Kansas to define a reference node. The intent of their research was to bring together the different Execution Environments existing at that time and enabled the definition of a standardized Execution Environment-NodeOS interface."

3.1.8 Active Network Architecture Open Issues

Proposition 8 *[Network Architecture Considerations]*
Many aspects of the global active network service seen by a user – the programming model, how packets are forwarded, how state is installed, how resources are reserved, etc. – are determined by the execution environment, and are thus beyond the scope of this document. This section deals with aspects of the network architecture that can or must be directly affected by the node architecture (in particular, the NodeOS). These aspects fall generally into the following categories:
- resource conservation mechanisms;
- information that must be interpreted by both user and node OS; and
- mechanisms that can be shared across Execution Environments for global efficiency.

The active network requires mechanisms to ensure that the network is robust, and to ensure that resource consumption by Execution Environments and users is controlled. In particular, it must not be possible for a malfunctioning or malicious Execution Environment or user to prevent other users from receiving adequate resources. Moreover, resource waste (e.g., due to routing loops) should be minimized. Execution Environments will play a role in implementing any such mechanisms, but the problem is complex, and it is possible that different end-to-end mechanisms will be appropriate for different Execution Environments. For this reason it is desirable to delineate the division of responsibility between the NodeOS and the Execution Environments with respect to limiting resource consumption. The primary resources of interest in the active network are transmission bandwidth, computing bandwidth, and storage. The first of these is arguably paramount – it is the reason for the existence of the network. However, in the active network, transmission bandwidth is available only via computation, and so the latter must also be protected. Storage is important both for transmission (buffering) and computation state. As the arbiter of access to individual node resources, the NodeOS must provide allocation and scheduling mechanisms to prevent users/Execution Environments from depriving other users/Execution Environments of access to those resources. These mechanisms must be fine-grained enough that an Execution Environment can rely on the NodeOS to protect individual users (of the same Execution Environment) from each other, and need not implement such mechanisms itself as another layer on top of the inter-Execution Environment isolation mechanisms. Thus, the NodeOS provides each Execution Environment (and potentially each user) with a 'controlled load' type of service: a virtual CPU/channel/memory that appears to be unloaded almost all of the time. The active network will provide for reservation of resources, and waste can occur if reserved resources are devoted exclusively to the reserving principal, i.e., are not available to others even when they are not being used by that principal. In order to preclude this possibility from the outset, the following fundamental requirement is imposed. Resource Conservation Rule 0: "All scheduling algorithms are work-conserving. Resources are also "wasted" when they are consumed by useless transmissions or computations. Obvious sources of these include malfunctioning Execution Environments or user programs; however, they may also arise from naturally occurring but transient conditions such as routing loops, network congestion, etc. It is desirable to bound effects of these and other failure modes. The classic example of a mechanism to bound is the IP TTL field, which bounds the number of times any individual packet will be transmitted in an IP network. The mechanism states that the TTL of a forwarded packet is one less than the TTL of the received packet. This mechanism requires state, but it is carried in the packet, and need not persist at any node. Unfortunately, in an active network the relationship between the packets received and transmitted by a node is at best unclear and at worst arbitrary. Therefore, the implementation of such mechanisms must be left to the individual Execution Environments."

Glaucon: "Why did the scheduling algorithms have to be work-conserving? I don't see how that prevents resource wastage."
Thrasymachus: "A work-conserving scheduling discipline implies that the server is never idle if packets await service. Thus, packets cannot accumulate in the queue waiting for service because the processor is idle or executing for some reason other than servicing incoming packets."

Socrates: "So are you saying that I cannot write code within an active packet that causes the active node scheduler to delay or obey my bidding in some other way?"

Thrasymachus: "Exactly. However, there is no reason that the code within your packet cannot delay itself within a particular node [2]. However, other packets continue to get service in the mean time."

Glaucon: "But isn't limiting resource consumption more of a global issue rather than specific to a single node?"

Thrasymachus: "This framework does not attempt to limit or even specify global mechanisms for resource usage; rather it attempts to focus on any specific resource limits that may be needed within a specific node."

Glaucon: "I am beginning to see the difference between 'bounding resource usage' and 'optimal resource allocation'. The former can be done locally while the latter must be done globally."

Thrasymachus: "But do not the two interact? Isn't 'bounding usage' part of 'optimal allocation'?"

Glaucon: "I suppose that is true. Then wouldn't a mechanism be necessary that allows a global allocation algorithm to have fine control over the bounding mechanism? Perhaps loosening one bound while tightening another?"

Thrasymachus: "This is a complex issue. Consider that we have not yet considered the 'value' of the active packets. The packet value can change over time [3]."

Glaucon: "How did the early prototype Execution Environments implement resource bounding?"

Socrates: "It could be implemented in many ways. For example, there was an Execution Environment called that embedded resource bounding in code written in a custom language called PLAN. PLAN programs had an associated finite resource bound that is decremented as the program runs. When the resource bound reaches zero, the program cannot execute any further. This prevents the program from running forever."

Glaucon: "Isn't that similar to the IP [4] Time To Live (TTL) field discussed in the document?"

Thrasymachus: "To an extent, yes. But the difference is that PLAN programs could spawn sub-programs, in which case the parent program grants a part or all of its resource bound to the children."

Socrates: "That explains monitoring during execution. But how did they assign the bounds in the first place? How did they decide on the units of measurement?"

Glaucon: "The next section outlines possible schemes for quantifying resource consumption."

3.1.9 Active Network Resource Control

Proposition 9 *[Quantifying Resource Needs]*
As discussed above, information about resource demands and usage needs to pass between the user and the active network. This information must quantify transmission, computation, and/or storage. The bit is universal as the quantitative measure of information transmission. Similarly, the byte is uniformly understood as a measure of storage capacity. For our purposes, it is probably a good idea to add a time dimension, to provide a bound on the effect of storing a byte. How to quantify computation requirements is less clear. Different active

nodes may have different underlying hardware architectures and different node operating systems, which impose different overheads. Moreover, the "program" whose computation is being described is Execution Environment-dependent, yet the quantity must be expressed in Execution Environment-independent terms. The computation involved in input and output channel processing must also be accounted for; it is not clear that the user will even be aware of the existence of such processing, much less how much it varies from node to node. One possibility is to use "RISC cycles" as the unit of computation, and to require that user-visible quantities always be given relative to the Execution Environment-specific code. (This might be user-supplied code, or it might be "standard" functionality built into the Execution Environment.) RISC cycles are essentially atomic, and it should be possible to define conversions for other kinds of hardware. Each active node would be responsible for translating "standard RISC cycles" to/from its own computing resource measurements, and accounting for all protocol processing on the input/output channels. For such a metric to be useful, the correspondence between the "source code" and the "object metric" of RISC cycles must be clear, or at least well-defined. In the case of standard code blocks, the code vendor would also supply the cycle requirements for specific classes of inputs to the code. For user-supplied code, the user might have to estimate, perhaps by referring to standard "benchmark" programs, whose requirements are published; alternatively, tools might be provided to automate the estimation based upon generic formulas. For all three resources – but especially for memory and computation – precision is likely to be limited. Moreover, the range of values to be specified is likely to be extensive. Therefore a logarithmic metric is recommended for measures of memory and computation.

Socrates: "Why does this part of the document exist? Certainly, as we have already discussed, communication networks have always performed some type of processing and high performance had always been a goal."

Glaucon: "Quantitative performance analysis techniques must have existed in some form prior to the existence of this document. Were not such techniques already used in building ancient non-active networks?"

Thrasymachus: "We must remember that the devices described in this document had never been open to developers to such a degree before. There is concern that proper resource allocation and bounding cannot be done without standard metrics for bandwidth, storage, and computation."

Socrates: "They appear to be most concerned about processing metrics."

Glaucon: "I suppose that is understandable, given the fact that these devices may run on any type of platform, with any processor, NodeOS, and Execution Environment."

Socrates: "Don't forget the input and output processing channels. These will add to the total amount of processing required."

Thrasymachus: "It appears that they are striving for a global standard metric for processing.[5] Each Execution Environment would then interpret that metric based upon knowledge of its own specific hard and software environment."

3.2 EXERCISES

1. If multiple EEs can reside on the same intermediate network node, how is the decision made as to which EE should receive a packet?
2. Describe the function of a channel in the Active Network Framework.
3. Explain how the current Internet Protocol can be implemented via an Active Network.
4. Explain how an Active Network can reside on top of the current Internet Protocol.
5. Generate a trace of an Active Network packet traveling from source to destination, labeling all events.
6. Explain at least four examples of composable services in detail. Explain the pros and cons of each.
7. Identify all instances of processing of an active packet within a typical active node.

Notes

[1]The characters in this dialog are from Plato's *The Republic*. The correlation between Plato's attempt to describe a perfect society in *The Republic* and the dialog in this book is only superficial. However, considering the network devices as a society, Plato's attempts to optimize resources, organize security of the state, and balance fairness to individuals with the good of society are somewhat analogous, with apology to Plato.

[2]In fact this was exactly what was done in order to implement an active jitter control mechanism [Bush et al., 2000].

[3]See [Hershey and Bush, 1999] for a discussion of active networks and the value to an application of message values that change over time.

[4]Internet Protocol

[5]The Anamo project at NIST is currently working on this. Further information can be found at http://w3.antd.nist.gov/activenets.

4

MANAGEMENT REFERENCE MODEL

This chapter discusses the goals and requirements for an active network management framework. The active network management framework refers to the minimum model that describes components and interactions necessary to support management within an active network. This is motivated by comparing management in current networks with the possibilities enabled to support management within active networks. Towards this objective, an overview of the current network management model is discussed as a prelude to discussing the active network management model.

In the current communications model, managed devices are viewed abstractly as protocol layer two and protocol layer three network devices that forward data from source towards destination end-systems. The actions taken by these devices are predefined and fixed for each protocol layer and packet type as shown in Figure 4.1. The figure shows non-active data packets transporting management requests to the managed device and a possible management response is shown leaving the managed device.

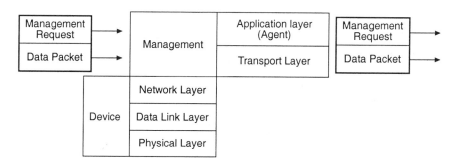

Figure 4.1. Current Management Model

The current management model, as illustrated and implemented by such protocols as the Simple Network Management Protocol (SNMP) and the Common Management Information Protocol (CMIP), requires that network devices have a management agent that responds to management requests. Devices must be addressable and respond directly to remote management commands. The model assumes that network nodes are instrumented with the ability to respond to requests for pre-configured data points of management information. Management information needs to be gathered for behavior of protocols in the higher layers of the stack, e.g., application data. This requires instrumenting more than just the bottom two or three standard protocol layers. Therefore, management has never been a natural fit to the current non-active communications model for intermediate network devices. It was initially considered difficult and uncommon for any type of standards-based management to exist because of the large number of non-interoperable proprietary attempts to solve the problem. Thus, the goal had been to implement a standard management framework that was robust and would be ubiquitously deployed across the Internet. The Simple Network Management Protocol had filled this role to some extent; however, active networks allow for a better solution.

In the current management model, shown in Figure 4.2, high-level queries are entered into, or generated from, a central management station that breaks the query down into low-level requests for data from managed entities. The current management model requires that all data values that would be needed for management must be predetermined and pre-defined in an information store called a Management Information Base (MIB). Each data point has a predetermined type, size, and access level and is called a Management Information Base Object. The result is that the Management Information Base contents, that is, the collection of Objects, must be painstakingly designed and agreed upon far ahead of time before they can be widely used. Even after accomplishing this, elements of the Management Information Base have static, inflexible types. This is antithetical to the objective of the active network framework, which seeks to minimize committee-based agreements. In an active network framework, elements of the Management Information Base have the potential to be dynamically defined and used by applications. The static data type of a Management Information Base may be reasonable for network hardware, but becomes less appropriate as higher layers of the protocol stack and applications are instrumented.

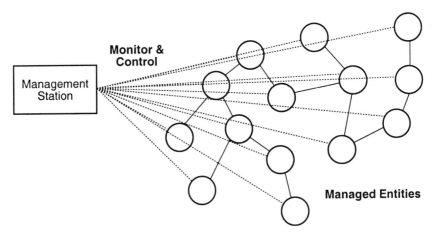

Figure 4.2. Current Centralized Management Model.

The current management model leads to a poor network control architecture. Large delays are incurred as agents send raw data to a central management station that takes time to refine and process the primitive data and perhaps respond with a Simple Network Management Protocol *Set Request* control action. However, the current management model has been primarily concerned with monitoring rather than control, in part because control has been hampered by the long transfer delay times to the centralized management station. While management *Set Requests* can be used for control purposes, few Management Information Bases today utilize the set commands for any type of real-time control.

As the Simple Network Management Protocol in current non-active networks has made steps toward providing reliable, integrated network management, the demand for more systems integrated management and control increases. Perhaps the demand is fed by the success of the Simple Network Management Protocol and by the explosion in the size of communication networks and number of applications utilizing them. Network administrators are pushing to extend network management ever higher towards and into the application layer. Integration is a primary driver. Clearly, applications, end-systems and the network all need to be managed in an integrated fashion.

The paradigm of instrumenting network elements is not the best solution for managing higher layer protocols and applications, especially in an active network wherein applications have a direct interaction with network elements. One reason is increased complexity. Network hardware devices and low-level network protocol layers behave in precise, well defined ways. On the other hand, active applications can interact with network protocols and other applications in myriad ways. This complexity in interaction requires a proportional increase in the number of management data points. Instrumenting every network device and end-system to support management of every application is not a scalable or feasible option. This could lead to other problems, such as redundant management data points, because two applications interacting with each other utilize the same management agent capability.

Another characteristic of the current management paradigm is that intermediate nodes are not designed to support management algorithms on the nodes themselves. However, fully integrated system management has always been the goal. Values from data points from all managed elements, including applications, are simply transported to a centralized management station where all refinement and processing takes place. Proxy agents are sometimes used to manage devices that have non-standard or non-existing management interfaces. Proxy agents serve as intermediary translators between the management standard and the operations of which the device is capable for management. The only other place that processing could be done within the network is in the managed object's agent. However, the prevailing philosophy has been that the managed object should be fully devoted to its primary task of forwarding data, not management, and therefore the agent is designed and implemented to be as simple and efficient a process as possible. The agent simply responds to requests for management data point values and generates unsolicited trap messages, hopefully, infrequently and only under extreme conditions.

Active networking affords an opportunity to take a new look at the network management problem and communications in general from a different perspective. It is a perspective that flips the traditional networking paradigm on its head. By allowing general-purpose computation on traditional intermediate network systems, it is no longer required that application processing, including network management processing, be restricted to end-systems. Optimum management efficiency can be achieved because processing can be allocated to intermediate network resources. This allows for a larger set of feasible solutions to the allocation of processing resources. For example, the old philosophy of keeping communications as simple possible has resulted in a plethora of highly specialized protocols illustrated by the large number of Internet Engineering Task Force Requests for Comments that are extant. In terms of network management, the old philosophy has caused enormous inefficiency by requiring large amounts of data to be transported to centralized management stations, even in instances when the data turns out to be of limited or no value. The active network model provides a communications model that is a better fit to the management model. In the active network reference model, intermediate-system active devices have the ability to accept and process any packet as a natural part of its packet processing, including network management packets. In fact, in the new management paradigm, management can be integrated into the processing framework itself; that is, packets are the application and manage themselves.

In the new management model shown in Figure 4.3, it is possible for a high-level management query in executable form to be sent directly to the managed active application. Because the managed application is active, it is implemented via active packets. The management query active packet interacts with the active application's packets in order to determine the result of the query. Given active network protocol composition, methods dynamically bound to an application no longer require a Management Information Base with static data point definitions to return predefined values, but instead, access local data points, compute a result from the local data and return only the final result, or some set of data culled from the local data that can lead to the final result, which may be computed in another part of the network. Many management systems today operate by polling a value and setting a threshold that trips an alarm when the threshold is crossed. Frequent queries result in wasted bandwidth if the threshold is rarely reached. The only information required in such cases is the alarm. In the active network management environment, the threshold crossing detection can be dynamically bound as a method in the managed active application.

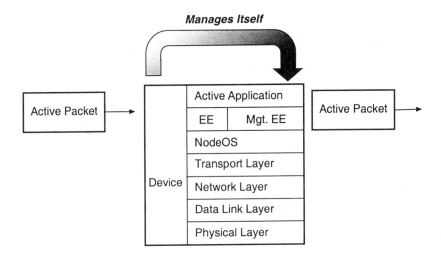

Figure 4.3. Active Management Model.

The old management philosophy requires that a *Set Request* be used for control purposes. This results in long delays when the controller is a centralized management station as compared to the active network model that enables local computation and control. Delays are clearly dangerous in a control system. Thus, using the Simple Network Management Protocol *Set Request* is also highly inefficient for dynamic control purposes. Active networks allow more distributed control for management purposes than in today's management model and an opportunity for a new management paradigm. The control algorithm is bound directly in the managed active application, thus reducing the delay incurred by dealing with a centralized management station. The active nature of the network also allows a framework in which efficient prediction of network behavior is possible. The Active Virtual Network Management Algorithm described in the next part of this book takes advantage of the active network to provide a model-based predictive management control framework. This requires a form of introspection that is possible in the new management model. Introspection is enabled because applications can control and manage themselves to a greater degree with active networks than ever before in the old management philosophy. The following example shows that data in an active network management model example has the ability to be queried by standards based network management protocols. A small agent is encapsulated with the active data as shown in Figure 4.5. When the data is queried, the agent responds with the values maintained by that specific agent's Management Information Base (MIB). The converse of this is shown in Figure 4.4, which illustrates active data containing a management client capable of querying management agents. This concept has been prototyped in Java in the active network testbed at General Electric Corporate Research and Development.

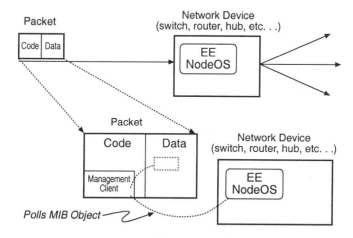

Figure 4.4. An Overview of the Traveling SNMP Client.

It has been recognized that, even in the new active management model, systems administrators require an integrated view of the entire managed system. However, note that integrated does not necessarily imply centralized. Also, note that the functionality of an integrated management view has changed dramatically from the old management model. The old model management view consisted of displaying values from static data types that are predefined. This is in contrast to the new that consists of controlling the algorithms (methods) that are bound to managed active entities and displaying results from those algorithms. Thus, the new management model deals with methods rather than data types.

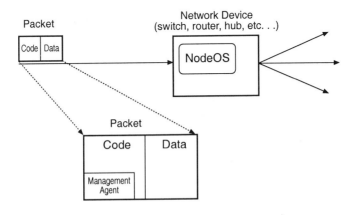

Figure 4.5. Queryable Data.

In the view of the authors, the new active network management goals should consist of:

- Automated self-management and control of applications.
- Ability to dynamically add/remove management features across all active applications.
- A richer integrated management view of the network and applications than in the old network management model.
- Decentralized and distributed management within the network for increased efficiency.
- Extreme reliability in the face of network failure.
- Support for integrated management of legacy applications

A few words of explanation are in order to justify why these goals are worthy of pursuit. Clearly, network management benefits from being as automated as possible. The words "self-management" are used because it is assumed that the system is able to determine best how to manage and control itself. An integrated view is the most concise and logical for human consumption and allows quick identification of correlated events. This assumes that a security policy mechanism is in place for network managers to gain access only to their own views of the system. The goal is that active networks will allow a richer semantic view of the integrated system. We want the system to be decentralized and distributed since that provides the most efficient use of resources and better response times. In addition, it can allow for graceful degradation of performance as resources fail. Finally, management is most critical when the system is failing. Thus the management system must be as robust and reliable as possible; that is, it should be the last service to fail. A framework within active networks that supports these goals is useful. However, care must be taken in developing a framework that does not preclude the development of general-purpose innovative management techniques enabled by active networking. The Active Virtual Network Management Prediction Algorithm is a step towards an active network management framework by enabling model-based predictive control, as discussed in the next section.

4.1 TOWARDS AN ACTIVE NETWORK MANAGEMENT FRAMEWORK

The previous section discussed the goals of a new framework for network management. Consider what is required from the framework in order to achieve these goals. Automated self-management and control of applications require application developers to provide monitoring and access into their applications. While an application may be self-managing and autonomous, it cannot be a completely closed system. The application needs information about other applications and the network that it resides upon. The application may need to negotiate with other applications for resources. The management interface between applications could be accomplished through definitions as is the case in today's non-active networks; however, more is possible with an active network. For example, the Management Information Base could itself become an active entity. Model-based predictive control is a particular mechanism enabled by the Active Virtual Network Management Prediction Algorithm described in detail in the next part of this book. A fully autonomous, self-managed application requires:
- Inter-application semantic specification
- Inner-loop control mechanisms
- Negotiation capability
- Managed data semantic correlation
- Security policy

The negotiation capability and inter-application semantic specification are of primary interest here because they require some form of semantic knowledge and goal seeking capability. While dealing with semantic knowledge and goal achieving research are major efforts in their own right, the new architecture should facilitate and encourage their development. The integrated management view requires that all the management information from each managed entity be brought together and presented to a single user. This means that a policy must be in place to control access to information and the data must have the ability to correlate itself with other data for an integrated view. This requires a security policy and managed data semantic correlation.

There are several spheres of management in the active network management model: the Execution Environment (EE), the Active Application (AA), and the Network management algorithm, where a network Management Application (MA) is a new management feature to be added to all Active Applications (AA). The ability to dynamically bind methods into active applications is an assumed feature in active networks. The actual mechanisms for inserting methods into an existing and executing application are discussed in (Zegura, 1998). A brief summary of example methods is presented in Table 4.1. Self-organizing management code, knowing when, where, and how to insert itself into the managed active application, is a goal that is partially met by the Active Virtual Network Management Prediction Algorithm. The Active Virtual Network Management Prediction framework discussed in detail in the next part of this book demonstrates the fundamental requirements of the new active network management framework, namely:

- Access to managed device monitoring and control.
- Insertion of monitoring elements into arbitrary locations of active applications.
- Injection of executable models onto managed nodes and/or into managed active applications.
- Injection/interception of management packets within the network.

Table 4.1. Active Network Composition Methods.

Composition Type	Reference
Functional	Hicks et al., 1999
Dataflow	Da Silva et al., 1998
Slots	Samrat Bhattercharjee, Kennth L. Calvert and Ellen W. Zegura, 1998
Signaling Extensions	Braden et al., 2000

4. 2 PREDICTION IN NETWORK MANAGEMENT

Network management is evolving from a static model of simply monitoring the state of the network to a more dynamic, feature-rich model that contains analysis, device and line utilization, and fault-finding capabilities. The management marketplace is rich in software to help monitor and analyze performance. However, a severe limitation of current state-of-the-art network management techniques is that they are inherently reactive. They attempt to isolate the problem and determine solutions after the problem has already occurred. An example of this situation is the denial of service attack on Internet portal Yahoo!'s servers on February 7, 2000. Network managers were only able to detect the attack and respond to it long after it crippled their servers. To prevent such occurrences, network management strategies have to be geared towards assessing and predicting potential problems based on current state. Another limitation of current management software is "effect-chasing." Effect chasing occurs when a problem causes a multitude of effects that management software misdiagnoses as causes themselves. Attempts to solve the causes instead of the problem result in wasted effort. Recent advances in network management tools have made use of artificial intelligence techniques for drilling down to the root cause of problems. Artificial Intelligence techniques sift through current data and use event correlation after the problem occurs to isolate the problem. While this provides a reasonable speedup in problem analysis, finding a solution can still be time-consuming because these tools require enough data to form their conclusions. Therefore, proactive management is a necessary ingredient for managing future networks. Part of the proactive capability is provided by analyzing current performance and predicting future performance based on likely future events and the network's reaction to those events. This can be a highly dynamic, computationally intensive operation. This has prevented management software from incorporating prediction capabilities. But distributed simulation techniques take advantage of parallel processing of information. If the management software can be distributed, it is possible to perform computation in parallel and aggregate the results to minimize computation overhead at each of the network nodes. Secondly, the usefulness of optimistic techniques has been well documented for improving the efficiency of simulations. In optimistic logical process synchronization techniques, also known as Time Warp (Bush et al., 1999; Bush, 1999), causality can be relaxed in order to trade model fidelity for speed. If the system that is being simulated can be queried in real time, prediction accuracy can be verified and measures taken to keep the simulation in line with actual performance. Networks present a highly dynamic environment in which new behaviors can be introduced as new applications inject new forms of data. The network management software would have to be highly adaptive to model these behaviors and analyze their effects.

Active networking provides an answer to this problem. Active networking offers a technology wherein applications can inject new protocols into the network for the network nodes to execute on behalf of the application. A network is defined to be an active network if it allows applications to inject customized programs into the network to modify the behavior of the network nodes. The nodes of the network, called active nodes, are programmable entities. Application code is embedded inside a special packet called a SmartPacket. When the SmartPacket reaches the appropriate active node, the code is extracted and executed at the node to implement new services. Active networking thus enables modification of a running simulation by injecting packets modeling the behavior of a new application into the network. This research presents a new proactive network management framework by combining the three key enabling technologies: (1) distributed simulation, (2) optimistic synchronization, and (3) active networking. The next section provides an introduction to the predictive framework and describes its various components.

4.2.1 Temporal Overlay

The approach taken by AVNMP is to inject an optimistic parallel distributed simulation of the network into the active network. This can be viewed as a virtual overlay network running temporally ahead of the actual network. A virtual network, representing the actual network, can be viewed as overlaying the actual network. A motivating factor for this approach is apparent when AVNMP is viewed as a model-based predictive control technique where the model resides inside the system to be controlled. The environment is an inherently parallel one; using a technique that takes maximum advantage of parallelism enhances the predictive capability. A well-known problem with parallel simulation is the blocking problem, in which processors are each driven by messages whose queues are attached to the processor. The message time-stamps are within the message. The message value is irrelevant. It is possible that one processor could execute a message with a given time stamp, then it could receive the next message with an earlier time-stamp. This is a violation of causality and could lead to an inaccurate result. There have been many proposed solutions to this problem. However, many solutions depend on the processor that is likely to receive messages out of order, waiting until the messages are guaranteed to arrive in the proper order. This increases delay and thus reduces the overall system performance. The AVNMP Algorithm makes use of a well-known optimistic approach that allows all processors to continue processing without delay, but with the possibility that a processor may have to rollback to a previous state. In addition the AVNMP Algorithm dynamically keeps the predictions within a given tolerance of actual values. Thus the model-based predictive system gains speedup due to parallelism while maintaining prediction accuracy.

The AVNMP system is comprised of Driving Processes, Logical Processes, and streptichrons, which are active virtual messages. The Logical Processes and Driving Processes execute within an Active Network Execution Environment (EE) on each active network node. The Logical Process manages the execution of the virtual overlay on a single node and is primarily responsible for handling rollback. Rollback can be induced by out-of-order Virtual Message arrivals and by prediction inaccuracy. A tolerance is set on the maximum allowable deviation between the predicted values and the actual values. If this tolerance is exceeded, a rollback to wallclock time occurs. The Logical Processes' notions of time only increment as virtual messages are executed. A sliding lookahead window is maintained so that a specified distance bounds the Logical Processes' virtual time progression into the future. The Driving Process monitors the input to that portion of the network enhanced by AVNMP and generates the Virtual Messages that drive the AVNMP Logical Processes forward in time. The driving process monitors the actual application via a general management frame developed within the active network environment. The driving process samples the values to be predicted and generates a prediction. The actual mechanism used for predicting output from any application is application dependent and de-coupled from the system. However, a simple curve-fitting algorithm based upon past history has worked adequately well.

4.2.2 Enhanced Message Capabilities

A Streptichron (from Classical Greek meaning to "bend time") is an active packet facilitating prediction that implements any of the active mechanisms described in this section. The streptichron can use this capability to refine its prediction as it travels through the network. In the initial AVNMP architecture, there was a one-to-one correspondence between virtual messages and real messages. While this correspondence works well for adding prediction to protocols using a relatively small portion of the total bandwidth, it is clearly beneficial to reduce message load, especially when attempting to add prediction of the bandwidth itself. There are more compact forms of representing future behavior within an active packet besides a virtual message. For relatively simple and easily modeled systems, only the model parameters need be sent and used as input to the logical process on the appropriate intermediate device. Note that this assumes that the intermediate network device's Logical Process is simulating the device operation and contains the appropriate model. However, because the payload of a virtual message is exactly the same as a real message, it can be passed to the actual device, and the result from the actual device is intercepted and cached. In this case, the Logical Process is a thin layer of code between the actual device and virtual messages primarily handling rollback. An entire executable model can be included within an active packet generated by the DP and executed by the Logical Process. When the active packet reaches the target device, the model provides virtual input messages to the Logical Process, and the payload of the virtual message is passed to the actual device as previously described. Autoanaplasis ("self adjust") is the self-adjusting characteristic of streptichrons. For example, in load prediction, streptichrons use the transit time to check prior predictions. General-purpose code contained within the packet is executed on intermediate nodes as the packet is forwarded to its destination. For example, a packet containing a prediction of traffic load may notice changes in traffic that influence the value it carries as the packet travels towards its destination. The active packet updates the prediction accordingly.

Time is critical in the architecture of the AVNMP Algorithm system; thus, most classes are derived from class Date. Class AvnmpTime handles relative time operations. Class Gvt uses the active GvtPackets class to calculate global virtual time. Class AvnmpLP handles the bulk of the processing including rollback. Class Driver generates and injects real and virtual messages into the system. The PP class either simulates or accesses an actual device on behalf of the Logical Process. The PP class may not need to simulate the device because the payload of a virtual message is exactly the same as a real message; thus, the payload of the virtual message can be passed to the actual device and the result from the actual device is intercepted and cached. In this case, the Logical Process is a thin layer of code between the actual device accessed by the PP class. The GvtPacket class implements the Global Virtual Time packet that is exchanged by all logical and driving processes to determine global virtual time. The AvnmpPacket class is derived from KU_SmartPacket_V2 and is the class from which GvtPacket and Streptichron classes are derived. Magician is a toolkit that provides a framework for creating SmartPackets as well as an environment for executing the SmartPackets. Magician is implemented in Java version 1.1. Version 1.1 was primarily chosen because it was the first version to support serialization. Serialization preserves the state of an object so that it can be transported or saved, and re-created at a later time. Therefore, in Magician, the executing entity is a Java object whose state is preserved as it traverses the active network. Magician adheres to the Active Network Encapsulation Protocol (ANEP) (Alexander et al., 1997) format when sending the Java class definitions and the Java object itself over the network. The details about the architecture of an active node in Magician and the exact format of a Magician SmartPacket are described in (Kulkarni et al., 1998). AVNMP runs as an active application (AA) inside the Magician environment. AVNMP queries Magician's state to perform resource monitoring and for load computation. Communication between different packets belonging to AVNMP and with other active applications like an SNMP-based real-time plotter takes place through smallstate, named caches that applications can create for storage, from which information can be retrieved. The remainder of this book discusses AVNMP and some of the surprising temporal complexities it introduces in greater levels of detail. While active networking provides the benefits previously discussed, it also adds to the complexity of the network. The additional complexity of active networks makes network and systems management a challenging and interesting problem because it is a problem in which distributed computing can now more easily be brought to bear because distributed computing algorithms can be more easily implemented and more quickly deployed in an active network. It will no longer suffice for network analysts to focus solely on traditional network performance characteristics such as load,delay, and throughput. Because active networking enables application computation to be performed within the network, the network performance must be optimized in tandem with applications. Delays through the network may be slightly longer because of computation, yet more work is done on behalf of the application. Thus metrics that include a closer association with applications are required. The next part of this book explains the design and development of active networks that are capable of predicting their own behavior and serve as a predictive active network management framework.

4.3 PREDICTIVE SYSTEMS DISCUSSION

Let us rollback (or had we advanced forward in time?) to our friends Socrates, Glaucon, and Thrasymachus as they turn towards a discussion of managing active networks.

Glauco: "It seems clear to me that to perform any type of prediction requires a projection forward in time at a rate faster than wallclock time. I suppose that a closed system, one that is totally self-contained, could be run forward in time very accurately. This is because it would have no interaction with elements running at wallclock time."

Thrasymachus: "I don't believe it. Even a completely closed system could exhibit chaotic behavior. And besides, even if a perfectly closed system existed, it would be of no use to anyone since we could not interact with it."

Socrates: "This sounds like an interesting topic. I am not as intelligent as either of you, so please help me follow where this discussion may lead. I believe, Glaucon, that you are searching for a simplified, ideal model in which to formulate predictive capability for an active network. Am I correct?"

Glaucon: "You are correct, Socrates."

Socrates: "We exist within the Universe and often attempt to predict events about ourselves within the Universe: weather, investments, political and military results -- consider the cleverly planned, but ill-fated attempts of Athens against Sparta.* We were part of that event, yet would have been hard pressed to have predicted its outcome. Is it better to be within the system or outside of the system for which you are attempting to compute a prediction?"

Thrasymachus: "Your question, Socrates, is a moot one. We can never truly be outside of a system and still interact with it. The mere act of measurement changes a system, however negligible. We can never know the truth. Even the supposed perfect abstraction that we use to model the world, Mathematics, cannot fully and completely describe itself, as Gödel has shown."

Glaucon: "Thrasymachus, do not be such a downer. We have come a long way in understanding the world around us. The scientific method of observation, hypothesis, and experimental validation continue to yield many new insights. Let us continue the quest for a predictive system, while realizing that perfection may not be possible in practice."

Socrates: "Well said, Glaucon. In fact, you have mentioned the scientific method. I think there is more to what you have said than you may realize. What is the fundamental activity in developing a hypothesis? Or, let me state it this way: How does one determine the best hypothesis if more than one appear equally valid in experimental validation?"

Glaucon: "One would prefer the simpler hypothesis. We seek to reduce complexity in our understanding of the world around us."

Socrates: "Excellent. How does one measure complexity?"

---

* This is the Pelopennesian War (431-404 B.C.) in which the defeat of the formerly liberal and free-thinking Athens by Sparta led to Athen's defeatist attitude and subsequent trial and execution of Socrates.

Thrasymachus: "Socrates, I believe I know where you heading with this line of reasoning...and it is pointless. Complexity is the size of the smallest algorithm or program that describes the information of which you wish to measure the complexity. However, this complexity is, in general, uncomputable. So again, you're leading us to a dead end as usual."

Glaucon: "Wait Thrasymachus, I wish to see where this would lead. What could this possibly have to do with Active Networks or predictive network management?"

Socrates: "What was the new feature that active networks had added to communication that never existed before?"

Glaucon: "Executable code within packets executed by intermediate nodes within the network."

Socrates: "Exactly. Active networks are much more amenable to algorithmic information. In other words, it becomes much easier to transmit algorithms than it ever had before active networks."

Thrasymachus: "Fine. I know where you are going here. You are going to say that we can now transmit executable models once, rather than passive data many times. But think of the overhead. What would you gain by transmitting a huge executable model of a system to a destination when it interacts only rarely with that destination?"

Glaucon: "I see your point Thrasymachus. We need to know when it is advantageous to transmit the model, and when to transmit only the passive data from that model. But how does all of this relate to predictive network management?"

Socrates: "In order to obtain predictive capability from an active network, we can inject a model of the network into the network itself. Sounds very Gödelian...if there is such a word."

Thrasymachus (sarcastic tone): "Very good. Now what about the effect that the model has upon the network? How can the model predict its own impact upon the network? Shall we inject a model of the model into the model? This is all nonsense. The system could never be perfectly accurate and the overhead would make it too slow."

Socrates: "Thrasymachus, is the complexity of a network node smaller than the length of the actual code on network node itself?"

Thrasymachus: "Unless the node and its code have been optimized to perfection, the complexity will be smaller. This is obvious."

Socrates: "Will the model injected into the network be more, or less complex than the node itself?"

Thrasymachus: "Less complex, Socrates. As we have already determined, the purpose of science is to find the least complex representation of a phenomenon. That is what a model represents."

Socrates: "Thrasymachus, will you agree that a communication network is by its very nature a highly distributed entity?"

Thrasymachus: "Clearly, the network is widely distributed."

Socrates: "Thus, an application that takes advantage of that large spatial area would benefit greatly, would it not?"

Thrasymachus: "Agreed."

Glaucon: "Are you suggesting, Socrates, that we use space to gain time in implementing our lower complexity models?"

Socrates: "Certainly that should allow the models injected into the network to project ahead of wallclock time."

Thrasymachus: "I see that you are attempting to trade off space, fidelity, and complexity in order to gain time, but this still sounds like a very tough problem and the devil will be in the details. Synchronization algorithms cannot gain the full processing power of all the processors in the distributed system. This is because messages must arrive in the proper order causing some parts of the system to slow down more than others waiting for messages to arrive in order."

Glaucon: "Optimistic distributed simulation algorithms do not slow down a priori. They assume messages arrive in the proper order and processing always continues full speed. If a message does arrive out of order at a processor, the processor must rollback to a previously known valid state, send out anti-messages to cancel the effects of now possibly invalid messages that it had sent, and continue processing from the rollback time incorporating the new message in its proper order."

Socrates: "If each processor executes at its own speed based upon its input messages, then each processor must have its own notion of time."

Glaucon: "That is correct. Each processor has its own Local Virtual Time."

Thrasymachus: "Let me understand this more concretely by a tangible analogy. Let us suppose that messages are ideas, processors are mind, and time is the advancement of knowledge. Each person advances his or her knowledge by listening to and combining ideas, thus generating new ideas for others to improve upon."

Socrates: "Very good. Now suppose one was to discover a previously unknown work by say.the philosopher Heraclitus. Suppose also that this work was so advanced for its time that it changed my thinking on previous work that I had done. I would need to go back to that previous work, remember what I had been thinking at that time, incorporate the new idea from Heraclitus, and generate a new result."

Thrasymachus: "But from society's perspective, this would not be enough. You would need to remember to whom you had communicated your previous ideas and give them the new result. This may cause those people, in turn, to modify their own past work."

Socrates: "Exactly. One can see the advancement of philosophy moving faster in some people and slower in others. The people in whom it moves slowest can impede the advancement for society in general. If the ideas (messages) could be transmitted and received in proper order of advancement among individuals, then progress by society would be fastest; rather than having to waste time and energy to go back and correct for new ideas."

Thrasymachus: "This sounds fantastic if the messages happen to arrive in causal order, that is, in the order in which they should be received. It also sounds terribly inefficient if messages arrive out-of-order."

Socrates: "Perhaps Complexity Theory can be of help here. It is known that the true measure of complexity of a string is reached when the program that describes the string is the smallest program that returns the string. As the program becomes smaller, it becomes more random. Thus, the program optimized for size is the more random program. Can this be true of time as well? Is the most compressed, thus most efficient, virtual time also the most random?"

Glaucon: "I am beginning to grasp what you are saying. If the rollbacks occur in random sequence, then perhaps the network is optimized; if there is any non-randomness, or pattern in the rollback sequence, then there is an opportunity to optimize the causality in some manner."

Thrasymachus (sarcastically): "Wonderful, another dead-end. There are no perfect tests for randomness. You can't even detect it, much less optimize it using this method."

Socrates: "Unfortunately, Thrasymachus, you are correct. If there were answers to the deep problems of randomness and complexity, and their relationship to time and space, these would result in great benefits to mankind."

The next part of this book attempts to address the concepts raised in this discussion. Chapters 5 and 6 discuss an implementation of the distributed network prediction framework that is included on the CD in this book. This framework enables the rollback mechanism explained by Socrates and Thrasymachus above. Chapter 7 discusses in detail the work on synchronization algorithms leading towards AVNMP. Chapter 8 builds the theory for relating performance, accuracy, and overhead of such a system. Chapter 9 considers many of Thrasymachus' arguments against the existence of such a predictive system.

4.4 EXERCISES

1. In a standard management environment in the current management model, what is a typical mechanism for measuring devices that have a non-standard management protocol or no management instrumentation at all?

2. Building a management protocol that allows a value to be set in current management systems in a non-trivial task. What problems are involved in handling a *Set Request* command in a best effort transport environment? How does this add to delay? Hint: Consider potential problems with synchronization.

3. How much bandwidth is wasted by calculating load at an interface in the old management model versus the new management model in which only the final result is returned?

4. How has the processing load been changed in the new network management framework compared to the current management framework?

5. How much bandwidth is saved if the new management model is used to insert a method that generates an alarm only when the load exceeds a certain threshold? Assume the current management model can only obtain the value via polling and the value exceeds the given threshold only 2% of the time. Assume that the variable changes at a maximum rate of 5 units per second and no error can be tolerated.

6. What is the expected delay to set a value that can be expected in the current management model given a 10Mb Ethernet with 100 management nodes each polling 10 values and one centralized management station? The decision to set a variable is based upon the results of the polled values. Assume the best case: there is no other traffic besides the management traffic. Also assume the management station processes all requests sequentially.

7. Compare and contrast the robustness and reliability of network management in the new model and the old model.

8. AVNMP was introduced at the end of this chapter. How can a managed object utilize prediction information, particularly with regards to control techniques?

9. Suppose that the management system, as well as the active packets themselves, begin to base routing decisions upon the expected future load of a link. What effect does that have on the predictive system? Hint: If routing decisions change, load most likely changes as a result.

Notes

[1][Bush et al., 1999] and [Bush, 2000] provide early thoughts on this concept.

III

AVNMP

5

AVNMP ARCHITECTURE

This chapter begins by describing the Active Virtual Network Management Prediction architecture and follows with an operational example. While the system attributes predicted by the Active Virtual Network Management Prediction Algorithm are generic, the focus of this book is load prediction. In the discussion that follows, new meaning is given to seemingly familiar terms from the area of parallel simulation. Terminology borrowed from previous distributed simulation algorithm descriptions has a slightly different meaning in Active Virtual Network Management Prediction; thus it is important that the terminology be precisely understood by the reader.

The Active Virtual Network Management Prediction Algorithm can be conceptualized as a model-based predictive control technique where the model resides inside the system being controlled. As shown in Figure 5.1, a virtual network representing the actual network can be viewed as overlaying the actual network. The system being controlled is a communications network comprised of many intermediate devices, each of which is an active network node. This is an inherently parallel system; the predictive capability is enhanced by using a technique that takes maximum advantage of parallelism.

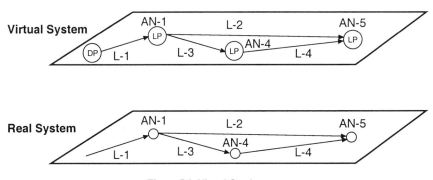

Figure 5.1. Virtual Overlay.

A well-known problem with parallel simulation is the blocking problem illustrated in Figure 5.2, where processors A, B, C, and D are each driven by messages whose queues are shown attached to the processor. The message time-stamps are indicated within the message. The message value is irrelevant. Notice that processor D could execute the message with time stamp 9, then it could receive the next message with time-stamp 6. This is a violation of causality and could lead to an inaccurate result. There have been many proposed solutions to this problem which are described in greater detail in following chapters of this book. However, many solutions depend on the processor that is likely to receive messages out of order waiting until the messages are guaranteed to arrive in the proper order. This adds delay and thus reduces the overall system performance. The Active Virtual Network Management Prediction Algorithm follows a well-known optimistic approach that allows all processors to continue processing without delay, but with the possibility that a processor may have to rollback to a previous state. In addition the Active Virtual Network Management Prediction Algorithm dynamically keeps the predictions within a given tolerance of actual values. Thus the model-based predictive system gains speed up due to parallelism while maintaining prediction accuracy.

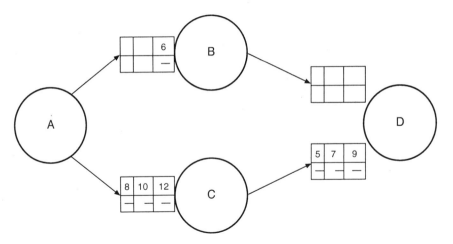

Figure 5.2. Blocked Process.

5. 1 AVNMP ARCHITECTURAL COMPONENTS

The Active Virtual Network Management Prediction algorithm encapsulates each Physical Process within a Logical Process. A Physical Process is nothing more than an executing task defined by program code. The Logical Process can be thinly designed to use the physical processes' software. If that is not possible, then the entire model can be designed into the Logical Process. An example of a Physical Process is the packet forwarding process on a router. A Logical Process consists of a Physical Process and additional data structures and instructions that maintain and correct operation as the system executes ahead of wallclock time as illustrated in Figure 5.3. As an example, the packet forwarding Physical Process is encapsulated in a Logical Process that maintains load values in its State Queue and handles rollback due to out-of-order input messages or out-of-tolerance real messages as explained later. A Logical Process contains a Send Queue (QS) and State Queue (SQ) within an active packet. In this implementation, the packet is encapsulated inside a Magician SmartPacket which follows the Active Network Encapsulation Protocol (Alexander et al., 1997) format. The Receive Queue maintains newly arriving messages in order by their Receive Time (TR). The Receive Queue is an object residing in an active node's smallstate. Smallstate is state left behind by an active packet. The Magician (Kulkarni et al., 1998) execution environment is used in the implementation described in this book. The Magician execution environment allows any kind of information to be stored in smallstate including Java objects; the Receive Queue is a Java object maintaining active virtual message ordering and scheduling. The Send Queue maintains copies of previously sent messages in order of their send times. The Send Queue is necessary for the generation of anti-messages for rollback described later. The state of a Logical Process is periodically saved in the State Queue. An important part of the architecture for network management is that the state queue of the Active Virtual Network Management Prediction system **is** the network Management Information Base. The Active Virtual Network Management Prediction values are the Simple Network Management Protocol Management Information Base Object values. They are the values expected to occur in the future. The current version of the Simple Network Management Protocol (Rose, 1991) has no mechanism for a managed object to report its future state; currently all results are reported assuming the state is valid at the current time. In working on predictive Active Network Management there is a need for managed entities to report their state information at times in the future. These times are unknown to the requester. A simple means to request and respond with future time information is to append the future time to all Management Information Base Object Identifiers that are predicted. This requires making these objects members of a table indexed by predicted time. Thus a Simple Network Management Protocol client that does not know the exact time of the next predicted value can issue a **get-next** command appending the current time to the known object identifier. The managed object responds with the requested object valid at the closest future time as shown in Figure 5.4.

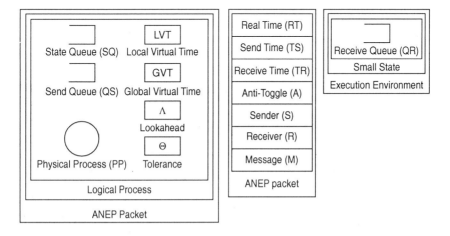

Figure 5.3. Active Global Virtual Time Calculation Overview.

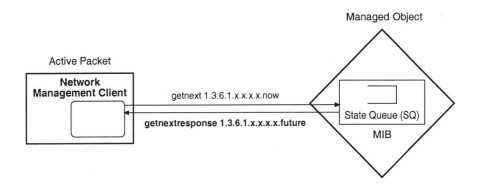

Figure 5.4. Legacy Network Management Future Time Request Mechanism.

The Logical Process also contains its notion of time, known as Local Virtual Time (LVT), and a Tolerance (Θ). Local Virtual Time advances to the of the next virtual message that is processed. Tolerance is the allowable deviation between actual and predicted values of incoming messages. For example, when a real message enters the load prediction Logical Process, the current load values are compared with the load values cached in the State Queue of the Logical Process. If predicted load values in the State Queue are out of tolerance, then corrective action is taken in the form of a rollback as explained later. Also, the Current State (CS) of a Logical Process is the current state of the structures and Physical Process encapsulated within a Logical Process.

5.1.1 Global Virtual Time

The Active Virtual Network Management Prediction system contains a notion of the complete system time known as Global Virtual Time (GVT) and a sliding window of length *Lookahead time* (Λ). Global Virtual Time is required primarily for the purpose of throttling forward prediction in Active Virtual Network Management Prediction; that is, it governs how far into the future the system predicts. There have been several proposals for efficient determination of Global Virtual Time, for example (Lazowaska and Lin, 1990) The algorithm in (Lazowaska and Lin, 1990) allows Global Virtual Time to be determined in a message-passing environment as opposed to the easier case of a shared memory environment. Active Virtual Network Management Prediction allows only message passing communication among Logical Processes. The algorithm in (Lazowaska and Lin, 1990) also allows normal processing to continue during the determination phase. A logical process that needs to determine the current Global Virtual Time does so by broadcasting a Global Virtual Time update request to all processes. Note that Global Virtual Time is the minimum of all logical process Local Virtual Times and the minimum message receive time that is in the system. An example is shown in Figure 5.5. The *Active Global Virtual Time Request Packet* notices that the logical process with a Global Virtual Time of 20 is greater than the last logical process that the *Active Global Virtual Time Request Packet* passed through and thus destroys itself. This limits unnecessary traffic and computation. The nodes that receive the *Active Global Virtual Time Request Packet* forward the result to the initiator of the Global Virtual Time request. As the *Active Global Virtual Time Request Packets* return to the initiator, the last packet is maintained in the cache of each logical process. If the value of the Active GVT Response is greater than or equal to the value in the cache, then the packet is dropped. Again, this reduces traffic and computation at the expense of space.

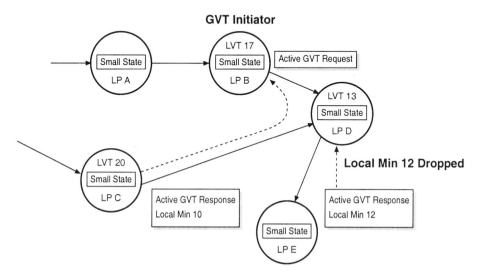

Figure 5.5. Active Global Virtual Time Calculation Overview.

5.1.2 AVNMP Message Structure

Active Virtual Network Management Prediction messages contain the Send Time (TS), Receive Time (TR), Anti-toggle (A) and the actual message object itself (M). The message is encapsulated in a Magician SmartPacket which follows the ANEP standard. The Receive Time is the time this message is predicted to be valid at the destination Logical Process. The Send Time is the time this message was sent by the originating Logical Process. The "A" field is the anti-toggle field and is used for creating an anti-message to remove the effects of false messages as described later. A message also contains a field for the current Real Time (RT). This is used to differentiate a real message from a virtual message. A message that is generated and time-stamped with the current time is called a real message. Messages that contain future event information and are time-stamped with a time greater than the current wallclock time are called virtual messages. If a message arrives at a Logical Process out of order or with invalid information, it is called a false message. A false message causes a Logical Process to rollback. The structures and message fields are shown in Table 5.1, Table 5.2 and in Figure 5.3. The Active Virtual Network Management Prediction algorithm requires a driving process to predict future events and inject them into the system. The driving process acts as a source of virtual messages for the Active Virtual Network Management Prediction system. All other processes react to virtual messages.

5.1.3 Rollback

A rollback is triggered either by messages arriving out of order at the Receive Queue of a Logical Process or by a predicted value previously computed by this Logical Process that is beyond the allowable tolerance. In either case, rollback is a mechanism by which a Logical Process returns to a known correct state. The rollback occurs in three phases. In the first phase, the state is restored to a time strictly earlier than the Receive Time of the false message. In the second phase, anti-messages are sent to cancel the effects of any invalid messages that had been generated before the arrival of the false message. An anti-message contains exactly the same contents as the original message with the exception of an anti-toggle bit which is set. When the anti-message and original message meet, they are both annihilated. The final phase consists of executing the Logical Process forward in time from its rollback state to the time the false message arrived. No messages are canceled or sent between the time to which the Logical Process rolled back and the time of the false message. These messages are correct; therefore, there is no need to cancel or re-send them, which improves performance and prevents additional rollbacks. Note that another false message or anti-message may arrive before this final phase has completed without causing problems. The Active Virtual Network Management Prediction Logical Process has the contents shown in Table 5.1, the message fields are shown in Table 5.2, and the message types are listed in Table 5.3 where t is the wallclock time at the receiving Logical Process.

Table 5.1. A VNMP Logical Process Structures

Structure	Description
Receive Queue (QR)	Ordered by message receive time (TR)
Send Queue (QS)	Ordered by message send time (TS)
Local Virtual Time	$LVT = \inf RQ$
Current State (CS)	State of the logical and physical process
State Queue (SQ)	States (CS) are periodically saved
Sliding Lookahead Window (SLW)	$SLW = (t, t + \Lambda)$
Tolerance (Θ)	Allowable deviation

Table 5.2 A VNMP Message Fields.

Field	Description
Send Time (TS)	LVT of sending process when message is sent
Receive Time (TR)	Scheduled time to be received by receiving process
Anti-toggle (A)	Identifies message as normal or antimessage
Message (M)	The actual contents of the message
Real Time (RT)	The wallclock time at which the message originated

Table 5.3 AVNMP Message Types.

Virtual Message	$RT > t$
Real Message	$RT \leq t$

5.1.4 Space-Time Trade-offs

The partitioning of physical processes into logical processes has an effect on the performance of the system. Active networks allow the possibility of physical processes to dynamically merge into logical process. In addition, both virtual and anti-messages can be fused on their way to their destination. There are several ways that this can occur. The first is a straightforward combination of data within the virtual messages when they reach a common node. Another fusion technique is to maintain a cache in each node of the last message that traveled through the node on the way to the message's destination for each source/destination pair. When a message arrives at a node to be forwarded towards its destination, it can check whether a message had been previously cached and if its Receive Time is greater than that of the current message. If so, this message knows it is going to cause a rollback. The message then checks whether it would have affected the result, for example, via a semantic check. If it would have had no effect, the message is discarded. In the specific case of load prediction, the change in load that the out-of-order message creates within the system can be easily checked. If many messages discover they would cause rollback on the way towards their destination, the destination logical process could perhaps be moved closer to the offending message generator logical process. If the message is a real message and the cached message is virtual and their times are not too far apart, a check can be made at that point as to whether a rollback is needed. If no rollback is needed, the real message can be dropped.

Virtual messages can be cached as they travel to their destination logical process. The cache uses a key consisting of the source-destination node of the message. Only the last message for that source-destination pair is cached. When the next message passes through the intermediate node matching that source-destination pair, the new message compares itself with the cached message. This is shown in Figure 5.6. If one exists and has a larger time-stamp, then a rollback is highly likely, and steps can be taken to mitigate the effects of the rollback. After the comparison, the old message is replaced in the cache with the new message. If many such rollback indications appear in the path of a virtual message, the destination process can be slowed or move itself to a new spatial location to mitigate the temporal effects of causality violations. Also, if a new message passing through an intermediate node is real, and the cached message is virtual, and they are within the same tolerance of time and value, the real message will destroy itself since it is redundant.

Logical Processes, because they are active packets, can move to locations that will improve performance. Logical Processes can even move between the network and end systems. In an extreme case of process migration, the Logical Processes are messages that install themselves only where needed to simulate a portion of the network as shown in Figure 5.7. Notice that choosing to simulate a single route always results in a feed-forward network.

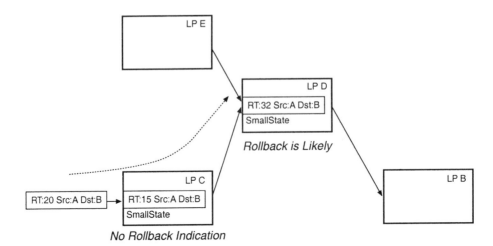

Figure 5.6. Active Rollback Mitigation.

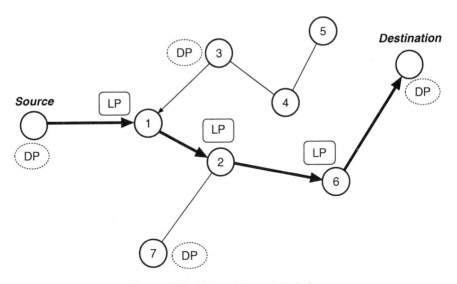

Figure 5.7. Partial Spatial Network Prediction.

5.1.5 Enhanced Message Capabilities

The active packet allows the virtual message to be enhanced with more processing capability. The virtual message can use this capability to refine its prediction as it travels through the network. In the Active Virtual Network Management Prediction architecture described thus far, there is a one-to-one correspondence between virtual messages and real messages. While this correspondence works well for adding prediction to protocols using a relatively small portion of the total bandwidth, it would clearly be beneficial to reduce message load, especially when attempting to add prediction of the bandwidth itself. There are more compact forms of representing future behavior within an active packet besides a virtual message. For relatively simple and easily modeled systems, only the model parameters need be sent and used as input to the logical process on the appropriate intermediate device. Note that this assumes that the intermediate network device's Logical Process is simulating the device operation and contains the appropriate model. However, because the payload of a virtual message is exactly the same as a real message, it can be passed to the actual device and the result from the actual device is intercepted and cached. In this case, the Logical Process is a thin layer of code between the actual device and virtual messages primarily handling rollback. An entire executable load model can be included within active packet generated by the DP and executed by the Logical Process. When the active packet reaches the target intermediate device, the load model provides virtual input messages to the and the payload of the virtual message passed to the actual device as previously described. A Streptichron is an active packet facilitating prediction as shown in Definition 5.1, which implements any of the above mechanisms.

$$\text{Streptichron} \triangleq \left\{ \begin{array}{l} \text{Input (Monte-Carlo) Model} \\ \text{Model Parameters (Self-Adjusting)} \\ \text{Virtual Message (Self-Adjusting)} \end{array} \right. \tag{5.1}$$

Autoanaplasis is the self-adjusting characteristic of streptichrons. For example, in load prediction, use the transit time to check prior predictions. Figure 5.8 shows an overview of autoanaplasis. General purpose code contained within the packet is executed on intermediate nodes as the packet is forwarded to its destination.

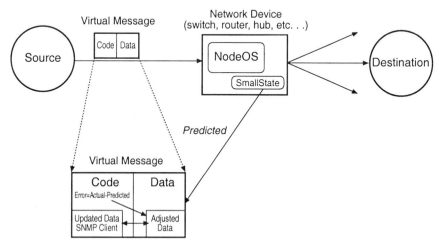

Figure 5.8. Self Adjusting Data.

For example, a packet containing a prediction of traffic load may notice changes in traffic that influence the value it carries as the packet travels towards its destination. The active packet updates the prediction accordingly.

5.1.6 Multiple Future Event Architecture

It is possible to anticipate alternative future events using a direct extension of the basic Active Virtual Network Management Prediction algorithm (Tinker and Agra, 1990). The driving process generates multiple virtual messages, one for each possible future event with corresponding probabilities of occurrence, or a ranking, for each event. Instead of a single Receive Queue for each Logical Process, multiple Receive Queues for each version of an event are created dynamically for each Logical Process. The logical process can dynamically create Receive Queues for each event and give priority to processing messages from the most likely versions' Receive Queues. This enhancement to Active Virtual Network Management Prediction has not been implemented. This architecture for implementing alternative futures, while a simple and natural extension of the Active Virtual Network Management Prediction algorithm, creates additional messages and increases the message sizes. Messages require an additional field to identify the probability of occurrence and an event identifier. Alternative future events can also be considered at a much lower level, in terms of perturbations in packet arrivals. Perturbation Analysis is described in more detail in (Ho, 1992).

5.1.7 Magician and AVNMP

The Active Virtual Network Management Prediction Algorithm has been built upon the Magician (Kulkarni et al., 1998) Execution Environment. This section discusses the development and architecture at the Execution Environment level. As discussed in the beginning of this book, Magician is a Java-based Execution Environment that was used to implement the Active Virtual Network Management Prediction Algorithm because at the time this project started, Magician had the greatest flexibility and capability. This included the ability to send active packets as Java objects. Figure 5.9 shows the Java class structure of the Active Virtual Network Management Prediction Algorithm implementation. Time is critical in the architecture of the system; thus, most classes are derived from class *Date*. Class *AvnmpTime* handles relative time operations. Class *Gvt* uses active the *GvtPackets* class to calculate global virtual time. Class *AvnmpLP* handles the bulk of the processing including rollback. Class *Driver* generates and injects real and virtual messages into the system. The *PP* class either simulates, or accesses, an actual device on behalf of the Logical Process. The *PP* class may not need to simulate the device because the payload of a virtual message is exactly the same as a real message; thus, the payload of the virtual message can be passed to the actual device and the result from the actual device is intercepted and cached. In this case, the Logical Process is a thin layer of code between the actual device accessed by the *PP* class. The *GvtPacket* class implements the Global Virtual Time packet which is exchanged by all logical and driving processes to determine global virtual time. Currently only the virtual message form of a streptichron has been implemented. The active packets have been implemented in both ANTS (Tennenhouse et al., 1997) and SmartPackets (Kulkarni et al., 1998).

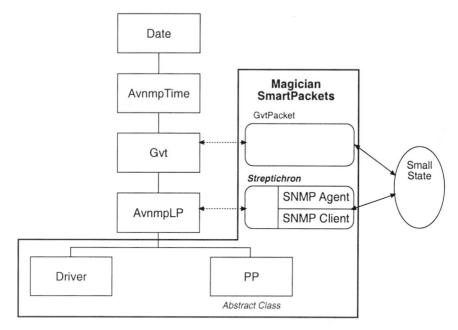

Figure 5.9. Active Virtual Network Management Protocol Class Hierarchy.

5.2 EXAMPLE DRIVING PROCESSES

5.2.1 Flow Prediction

Network flows are comprised of streams of packets. The ultimate goal for network management of flows is to allocate resources in order to provide the best quality of service possible for all user flows within the network. However, knowledge of how best to allocate resources is greatly aided by knowledge of future usage. Active Virtual Network Management Prediction provides that future usage information. The Active Virtual Network Management Prediction driving processes generate virtual load messages. The manner in which the prediction is accomplished is irrelevant to Active Virtual Network Management Prediction. Some example techniques could include a Wavelet-based technique described in (Ma and Ji, 1998) or simple regression models (Pandit and Wu, 1983).

5.2.2 Mobility Prediction

Proposed mobile networking architectures and protocols involve predictive mobility management schemes. For example, an optimization to a Mobile IP-like protocol using IP-Multicast is described in (Seshan et al., 1996). Hand-offs are anticipated and data is multicast to nodes within the neighborhood of the predicted handoff. These nodes intelligently buffer the data so that no matter where the mobile host (MH) re-associates after a handoff, no data will be lost. Another example (Liu et al., 1995) (Liu, 1996) proposes deploying mobile floating agents, which decouple services and resources from the underlying network. These agents would be pre-assigned and pre-connected to predicted user locations.

The Active Virtual Network Management Prediction driving process for mobile systems requires accurate position prediction. A non-active form of Active Virtual Network Management Prediction has been used for a rapidly deployable wireless mobile network as described in (Bush, 1997). Previous mobile host location prediction algorithms have focused on an aggregate view of mobile host location prediction, primarily for such purposes as base-station channel assignment and base-station capacity planning. Examples are a fluid flow model (Thomas et al., 1988) and the method of Hong and Rappaport (Hong and Rappaport, 1986). A location prediction algorithm accurate enough for individual mobile host prediction has been developed in (Liu and Jr., 1995). A brief overview of the algorithm follows because the algorithm in (Liu and Jr., 1995) is an ideal example of a driving process for Active Virtual Network Management Prediction and demonstrates the speedup that Active Virtual Network Management Prediction is capable of providing with this prediction method. The algorithm allows individual mobile hosts to predict their future movement based on past history and known constraints in the mobile host's path.

All movement ($\{M(k,t)\}$) is broken into two parts, regular and random motion. A Markov model is formed based on past history of regular and random motion and used to build a prediction mechanism for future movement as shown in Equation 6.1. The regular movement is identified by $S_{k,t}$ where S is the state (geographical cell area) identified by state index k at time t and the random movement is identified similarly by $X(k,t)$. $M(k,t)$ is the sum of the regular and random movement.

$$\{M(k,t)\} = \{S_{k,t} \mid k \leq K, t \in T\} + \{X(k,t) \mid k \leq K, t \in T\} \qquad (6.1)$$

$$\{X(k,t)\} = \{M(k,t)\} - (\{M_c(k,t) \mid k \leq K, t \in T\} + \{M_t(k,t) \mid k \leq K, t \in T\}) \qquad (6.2)$$

The mobile host location prediction algorithm in (Liu and Jr., 1995) determines regular movement as it occurs, then classifies and saves each regular move as part of a movement track or movement circle. A movement circle is a series of position states that lead back to the initial state, while a movement track leads from one point to another distinct point. A movement circle can be composed of movement tracks. Let M_c denote a movement circle and M_t denote a movement track. Then Equation 6.2 shows the random portion of the movement.

The result of this algorithm is a constantly updating model of past movement classified into regular and random movement. The proportion of random movement to regular movement is called the randomness factor. Simulation of this mobility algorithm in (Liu and Jr., 1995) indicates a prediction efficiency of 95%. The prediction efficiency is defined as the rate over the regularity factor. The prediction accuracy rate is defined in (Liu and Jr., 1995) as the probability of a correct prediction. The regularity factor is the proportion of regular states, $\{S_{k,t}\}$, to random states $\{X(k,t)\}$. The theoretically optimum line in (Liu and Jr., 1995, p. 143) may have been better labeled the deterministic line. The deterministic line is an upper bound on prediction performance for all *regular* movement. The addition of the random portion of the movement may increase or decrease actual prediction results above or below the deterministic line. A theoretically optimum (deterministic) prediction accuracy rate is one with a randomness factor of zero and a regularity factor of one. The algorithm in (Liu and Jr., 1995) does slightly worse than expected for completely deterministic regular movement, but it improves as movement becomes more random. As a prediction algorithm for Active Virtual Network Management Prediction, a state as defined in (Liu and Jr., 1995) is chosen such that the area of the state corresponds exactly to the Active Virtual Network Management Prediction tolerance, then based on the prediction accuracy rate in the graph shown in (Liu and Jr., 1995, p. 143) the probability of being out of tolerance is less than 30% if the random movement ratio is kept below 0.4. An out-of-tolerance proportion of less than 30% where virtual messages are transmitted at a rate of $\lambda_{vm} = 0.03$ per millisecond results in a significant speedup as shown in Chapter 8.

5.2.3 Vulnerability Prediction

Network vulnerability to information warfare attack can be quantified and vulnerability paths through the network can be identified. General Electric Corporate Research & Development has a patent disclosure on such a system. The results of this vulnerability system are used to identify the most likely path of an attack, thus predicting the next move of a knowledgeable attacker.

Once an attack has been detected, the network command and control center can respond to the attack by repositioning safe-guards and by modifying services used by the attacker. However, cutting-off services to the attacker also impacts legitimate network users, and a careful balance must be maintained between minimizing the threat from the attack and maximizing service to customers. For example, various stages of an attack are shown in Figure 5.10. Since the allocation of resources never changes throughout the attack in this specific scenario, the vulnerability of the target increases significantly with each step of the attack.

A probabilistic and maximum flow analysis technique for quantifying network vulnerability have been developed at General Electric Corporate Research & Development (Bush and Barnett, 1998). The results from that work are the probability of an attacker advancing through multiple vulnerabilities and the maximum flow or rate. Using this information, the logical processes in Figure 5.11 can predict when and where the attacker is likely to proceed and can update the graphical interface with this information before the attack is successful. This allows time for various countermeasures to be taken or the opportunity to open an easier path for the attacker to a "fish bowl," a portion of the network where attackers are unknowingly steered in order to watch their activity. Virtual messages are exchanged between the Information Warfare Command and Control and the logical processes in Figure 5.11.

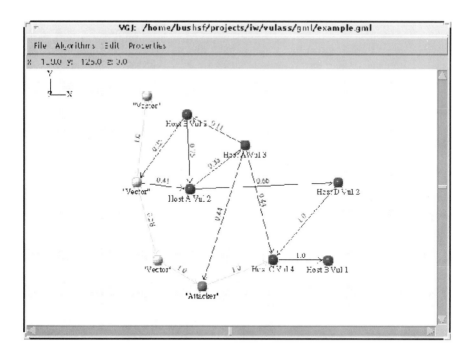

Figure 5.10. An Example of an Attack in Progress.

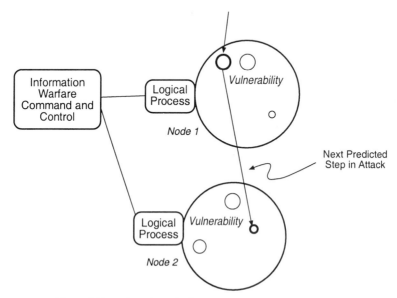

Figure 5.11. An Overview of Information Warfare Attack Prediction.

5.3 EXERCISES

1. Why is the send queue required in the AVNMP architecture?
2. How does the rollback mechanism work? Describe at least two methods to minimize rollback overhead.
3. In what ways does the Active Network facilitate the implementation of AVNMP? Could autoanaplasis exist in a non-active network?
4. Identify the two main causes of rollback and describe the relative impact of each cause on overhead.
5. Describe in detail the effect on the AVNMP system of unsynchronized real-time (wallclock) clocks at each node.
6. Describe a fossil collection algorithm that keeps all the queues in the AVNMP architecture as small as possible.

6

AVNMP OPERATIONAL EXAMPLES

The driving processes can make local predictions about load, vulnerability (Bush and Barnett, 1998), and mobile location (Bush, 1997). Load can be used to predict local QoS, congestion, and faults. The focus of this book is on the development and application of the Active Virtual Network Management Prediction algorithm and not the predictive methods within the driving processes. The primary purpose of Active Virtual Network Management Prediction is to distribute local changes throughout the network in both space and time.

Various predictive techniques can be used such as regression-based methods based on past history or similar techniques in the Wavelet domain. Since the Active Virtual Network Management Prediction implementation follows good modular programming style, the driving process has been decoupled from the actual prediction algorithm. Active Virtual Network Management Prediction has been tested by executing it in a situation where the outputs and internal state are known ahead of time as a function of the driving process prediction. The prediction within the driving processes is then corrupted and the Active Virtual Network Management Prediction output examined to determine the effect of the incorrect predictions on the system.

6.1 AVNMP OPERATIONAL EXAMPLE

A specific operational example of the Active Virtual Network Management Prediction Algorithm used for load prediction and management is shown in Figures 6.2 through 6.10. This particular execution log is from the operation of Active Virtual Network Management Prediction running on a simple three node network with an active end-system and two active intermediate nodes: AH-1, AN-1, AN-2. The legend used to indicate Active Virtual Network Management Prediction events is shown in Figure 6.1.

The Active Virtual Network Management Prediction system illustrated throughout this book has been developed using the Magician (Kulkarni et al., 1998) active network execution environment; the driving processes, logical processes, and virtual messages are implemented as Magician Smartpackets.

Figure 6.1. Legend of Operational Events.

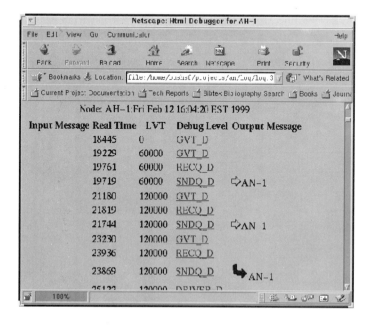

Figure 6.2. Active Node AH-1 Driving Process.

The logical process and driving process are injected into the network. The logical process automatically spawns copies of itself onto intermediate nodes within the network while the driving processes migrate to end-systems and begin taking load measurements in order to predict load and inject virtual messages. At the start of the Logical Process's execution, Local Virtual Time and Global Virtual Time are set to zero, lookahead (Λ) is set to 60,000 microseconds, and a (Θ) of 1,000 bytes/second is allowed between predicted and actual values for this process.

The description of the algorithm begins with an Active Virtual Network Management Prediction enabled network that has just been turned on and is generating real messages. The real messages in this case are randomly generated Magician Smartpackets running over a local area network. The driving process, is located on active node, AH-1. The driving process generates predictions about usage in the near future and injects virtual messages based on those predictions as shown in Figure 6.2. Figure 6.2 illustrates the log format used for the top-level view of all the Active Virtual Network Management Prediction Logical Processes. The left-most column shows incoming messages, the next column shows the wallclock time in microseconds, the next column shows the Local Virtual Time, the next column is a link to more detailed information about the event, and the right-most column shows any output messages that are generated. Both the input and output messages indicate the type of message by the legend shown in Figure 6.1 and are labeled with the source or destination of the message. Active node AH-1 shows two virtual active packets and one real active packet sent to AN-1.

6.1.1 Normal Operation Example

In Figure 6.3, active node AN-1 has begun running and receives the first virtual message from AH-1. AN-1's Logical Process must first determine whether it is virtual or real by examining the field. If the active packet is a virtual active packet, the Logical Process compares the message with its Local Virtual Time to determine whether a rollback is necessary due to an out-of-order message. If the message has not arrived in the past relative to the Logical Process's Virtual Time, the message then enters the Receive Queue in order by Receive Time. The Logical Process takes the next message from the Receive Queue, updates its Local Virtual Time, and processes the message (shown below the current view in Figure 6.3. Figure 6.4 shows the AN-1 state after receiving the first virtual message.

If an outgoing message is generated, as shown in Figure 6.5, a copy of the message is saved in the State Queue, the Receive Time is set, and the Send Time is set to the current Local Virtual Time. The message is then sent to the destination Logical Process. If the virtual message arrived out of order, the Logical Process must rollback as described in the previous section. Figure 6.6 shows AN-1's Local Virtual Time, Send Queue contents, contents, and contents after the received virtual message has been processed and forwarded. Figure 6.7 shows AN-1's state after sending the first virtual message.

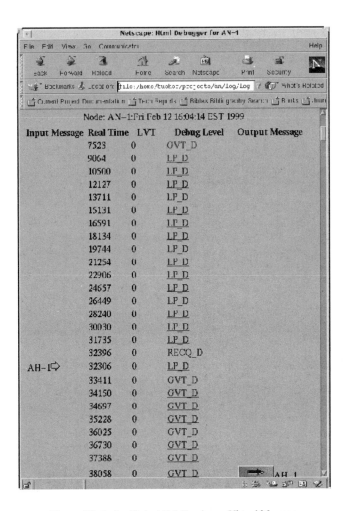

Figure 6.3. Active Node AN-1 Receives a Virtual Message.

6.1.2 Out-of-Tolerance Rollback Example

An example of out of tolerance rollback is illustrated in Figure 6.8. A real message arrives and its message contents are compared with the closest saved state value. The message value is out of tolerance; therefore, all state queue values with times greater than the receive time of the real message are discarded.

The send queue message anti-toggle is set and the anti-message is sent. The invalid states are discarded. The rollback causes the Logical Process to go back to time 120000 because that is the time of the most recent saved state that is less than the time of the out-of-tolerance message's Receive Time.

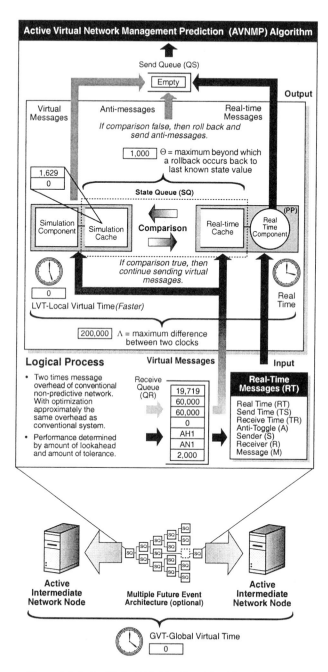

Figure 6.4. Active Node AN-1 After Receiving Virtual Message.

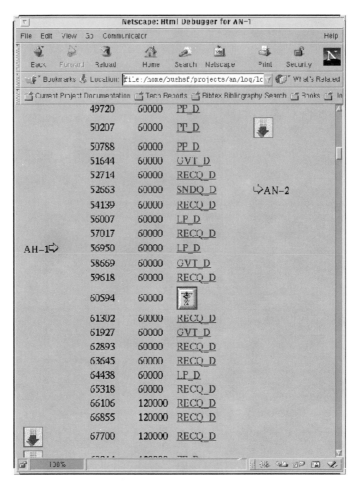

Figure 6.5. Active Node AN-1 Sends a Virtual Message.

Figure 6.12 shows the first virtual message received by AN-2. Figure 6.11 shows the AN-1 state after the first rollback. The anti-messages are the messages in the Send Queue that are crossed out. When these messages are sent as anti-messages, the anti-toggle bit is set. Also shown in Figure 6.11 is the discarded State Queue element that is no longer valid.

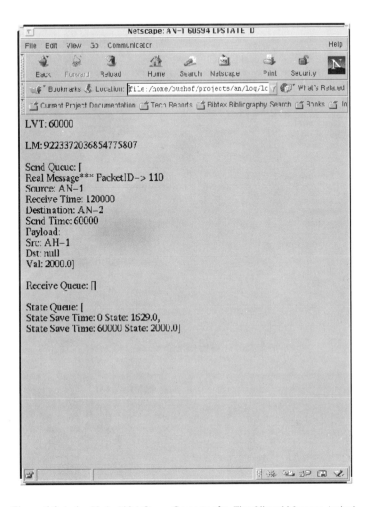

Figure 6.6. Active Node AN-1 Queue Contents after First Virtual Message Arrival.

6.1.3 Example Performance Results

Figure 6.13 shows the Local Virtual Time of node AN-1 versus wallclock time. Note that the logical process on AN-1 quickly predicted load 200,000 milliseconds ahead of wallclock time and then maintained the 200,000 millisecond lookahead. The sudden downward spikes in the plot are rollbacks.

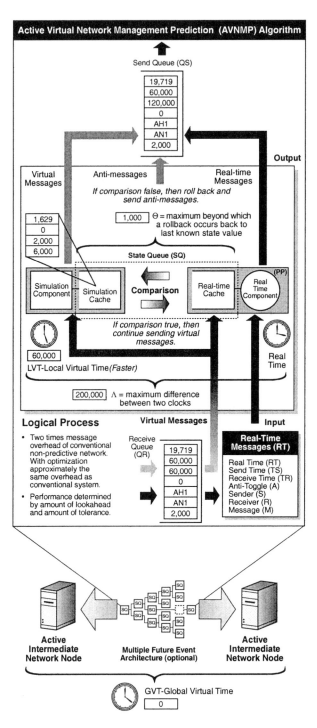

Figure 6.7. Active Node AN-1 after Sending Virtual Message.

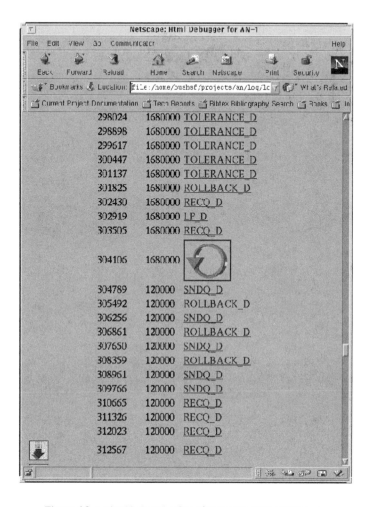

Figure 6.8. Active Node AN-1 Out-of-Tolerance Rollback Occurs.

A more complete view can be seen in the three-dimensional graph of Figure 6.14. The predicted values are shown as a function of wallclock time and LVT. This data was collected by SNMP polling an active execution environment that was enhanced with AVNMP. The valleys between the peaks are caused by the polling delay. A diagonal line on the LVT/Wallclock plane from the front right corner to the back left corner separates LVT in the past from LVT in the future; future LVT is towards the back of the graph, past LVT is in the front of the graph. Starting from the front, right hand corner, examine slices of fixed wallclock time over LVT; this shows both the past values and the predicted value for that fixed wallclock time.

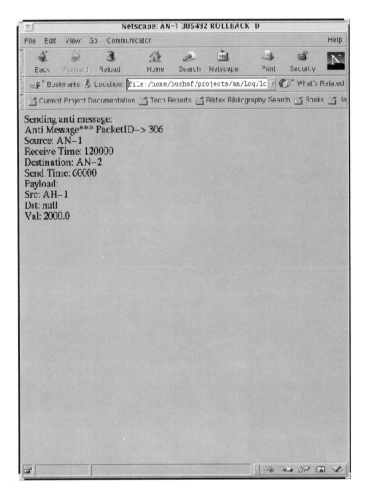

Figure 6.9. Active Node AN-1 Anti-Message Sent after First Rollback.

As wallclock time progresses, the system corrects for out-of-tolerance predictions. Thus, LVT values in the past relative to wallclock are corrected. By examining a fixed LVT slice, the prediction accuracy can be determined from the graph.

This chapter described the architecture and operation of the Active Virtual Network Management Prediction Algorithm. The performance of the algorithm is impacted by the accuracy of the predictions generated by the driving processes. The architecture is execution environment independent; however, the implementation used Magician. The next section discusses the driving processes in more detail. The remaining chapters of the book include analysis of the effect upon the system of driving process parameters such as virtual message generation rate, the ratio of virtual to real messages, and the prediction stepsize.

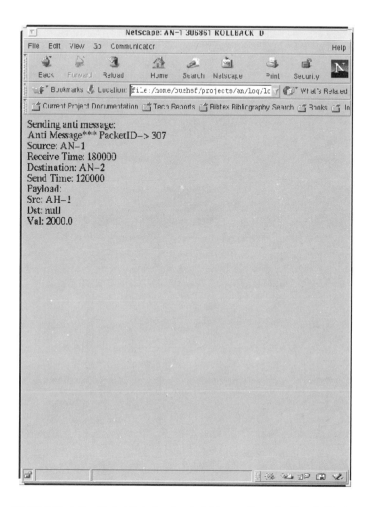

Figure 6.10. Active Node AN-1; Another Anti-Message Sent after First Rollback.

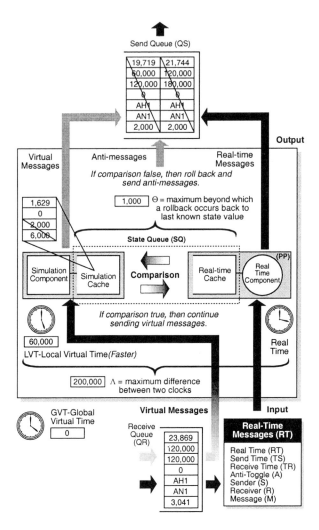

Figure 6.11. Active Node AN-1 after Rollback.

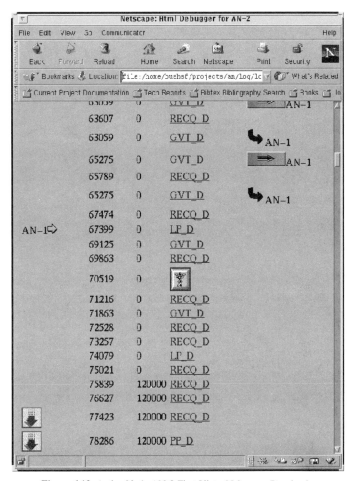

Figure 6.12. Active Node AN-2 First Virtual Message Received.

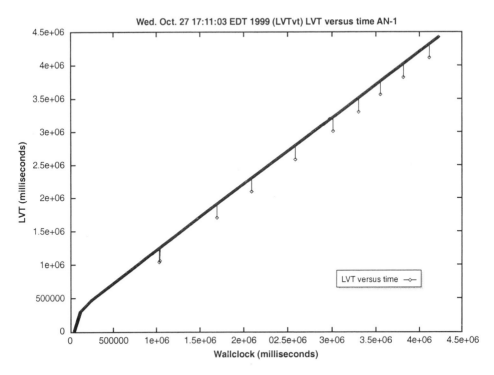

Figure 6.13. Active Node AN-1 LVT versus Wallclock.

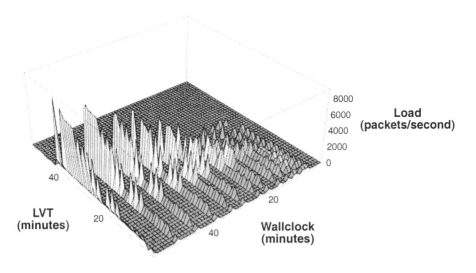

Figure 6.14. Three-Dimensional Graph Illustrating Predicted Load Values as a Function of Wallclock Time and LVT.

6.2 EXERCISES

1. Explain the impact of inaccurate prediction of the Driving Processes upon the AVNMP system.
2. Explain how multiple futures in AVNMP and vulnerability prediction could work together.
3. Explain how AVNMP prediction and intrusion detection could work together.
4. In what ways could AVNMP benefit a wireless system requiring setup time for a handoff from one base station to another?
5. Identify as many parameters as possible that affect the ability of AVNMP to maintain predictions as close to the maximum lookahead as possible.
6. Why did the two messages in the send queue disappear in Figure 6.11? What happened to them?
7. Notice the downward pointing spikes in Figure 6.13. What type of rollback do you think caused these spikes and why?

7

AVNMP ALGORITHM DESCRIPTION

One of the major contributions of this research is to recognize and define an entirely new branch of the Time Warp Family Tree of algorithms. Active Virtual Network Management Prediction integrates real and virtual time at a fundamental level allowing processes to execute ahead in time. The Active Virtual Network Management Prediction algorithm must run in real-time, that is, with hard real-time constraints.

7.1 FUNDAMENTALS OF DISTRIBUTED SIMULATION

Consider the work leading towards the predictive Active Virtual Network Management Prediction algorithm starting from a classic paper on synchronizing clocks in a distributed environment (Lamport, 1978). A theorem from this paper limits the amount of parallelism in any distributed simulation algorithm:

Rule 1: *If two events are scheduled for the same process, then the event with the smaller timestamp must be executed before the one with the larger timestamp.*

Rule 2: *If an event executed at a process results in the scheduling of another event at a different process, then the former must be executed before the latter.*

A parallel simulation method, known as CMB (Chandy-Misra-Bryant), that predates Time Warp (Jefferson and Sowizral, 1982) is described in (Chandy and Misra, 1979). CMB is a conservative algorithm that uses Null Messages to preserve message order and avoid deadlock. Another method developed by the same author does not require Null Message overhead, but includes a central controller to maintain consistency and detect and break deadlock. There has been much research towards finding a faster algorithm, and many algorithms claiming to be faster have compared themselves against the CMB method.

7.2 BASICS OF OPTIMISTIC SIMULATION

The basic Time Warp Algorithm (Jefferson and Sowizral, 1982) was a major advance in distributed simulation. Time Warp is an algorithm used to speedup Parallel Discrete Event Simulation by taking advantage of parallelism among multiple processors. It is an optimistic method because all messages are assumed to arrive in order and are processed as soon as possible. If a message arrives out-of-order at a Logical Process, the Logical Process rolls back to a state that was saved prior to the arrival of the out-of-order message. Rollback occurs by sending copies of all previously generated messages as anti-messages. Anti-messages are exact copies of the original message, except and anti-bit is set within the field of the message. When the anti-message and real message meet, both messages are removed. Thus, the rollback cancels the effects of out-of-order messages. The rollback mechanism is a key part of Active Virtual Network Management Prediction, and algorithms that improve Time Warp and rollback also improve Active Virtual Network Management Prediction. There continues to be an explosion of new ideas and protocols for improving Time Warp. An advantage to using a Time Warp based algorithm is the ability to leverage future optimizations. There have been many variations and improvements to this basic algorithm for parallel simulation. A collection of optimizations to Time Warp is provided in (Fujimoto, 1990). The technical report describing Time Warp (Jefferson and Sowizral, 1982) does not solve the problem of determining Global Virtual Time; however, an efficient algorithm for the determination of Global Virtual Time is presented in (Lazowaska and Lin, 1990). This algorithm does not require message acknowledgments, thus increasing the performance, yet the algorithm works with unreliable communication links.

An analytical comparison of CMB and Time Warp is the focus of (Lin and Lazowska, 1990). In this paper the comparison is done for the simplified case of feed-forward and feedback networks. Conditions are developed for Time Warp to be conservative optimal. Conservative optimal means that the time to complete a simulation is less than or equal to the critical path (Berry and Jefferson, 1985) through the event-precedence graph of a simulation.

7.3 ANALYSIS OF OPTIMISTIC SIMULATION

A search for the upper bound of the performance of Time Warp versus synchronous distributed processing methods is presented in (Felderman and Kleinrock, 1990). Both methods are analyzed in a feed-forward network with exponential processing times for each task. The analysis in (Felderman and Kleinrock, 1990) assumes that no Time Warp optimizations are used. The result is that Time Warp has an expected potential speedup of no more than the natural logarithm of P over the synchronous method where P is the number of processors.

A Markov Chain analysis model of Time Warp is given in (Gupta et al., 1991). This analysis uses standard exponential simplifying assumptions to obtain closed form results for performance measures such as the fraction of processed events that commit, speedup, rollback recovery, expected length of rollback, probability mass function for the number of uncommitted processed events, probability distribution function of the local virtual time of a process, and the fraction of time the processors remain idle. Although the analysis appears to be the most comprehensive analysis to date, it has many simplifying assumptions such as no communications delay, unbounded buffers, constant message population, message destinations are uniformly distributed, and rollback takes no time.

Thus, the analysis in (Gupta et al., 1991) is not directly applicable to the time sensitive nature of Active Virtual Network Management Prediction.

Further proof that Time Warp out-performs is provided in (Lipton and Mizell, 1990). This is done by showing that there exists a simulation model that out-performs CMB by exactly the number of processors used, but that no such model in which CMB out-performs Time Warp by a factor of the number of processors used exists.

A detailed comparison of the CMB and Time Warp methods is presented in (Lin, 1990). It is shown that Time Warp out-performs conservative methods under most conditions. Improvements to Time Warp are suggested by reducing the overhead of state saving information and the introduction of a global virtual time calculation. Simulation study results of Time Warp are presented in(Turnbull, 1992). Various parameters such as communication delay, process delay, and process topology are varied, and conditions under which Time Warp and CMB perform best are determined.

The major contribution of this section is to recognize and define an entirely new branch of the Time Warp Family Tree of algorithms, shown in Figure 7.1, that integrates real and virtual time at a fundamental level. The Active Virtual Network Management Prediction algorithm must run in real-time, that is, with hard real-time constraints. Real-time constraints for a time warp simulation system are discussed in (Ghosh et al., 1993). The focus in (Ghosh et al., 1993) is the R-Schedulability of events in Time Warp. Each event is assigned a real-time deadline $(d_{E_i,T})$ for its execution in the simulation. R-Schedulability means that there exists a finite value (R) such that if each event's execution time is increased by R, the event can still be completed before its deadline. The first theorem from (Ghosh et al., 1993) is that if there is no constraint on the number of such false events that may be created between any two successive true events on a Logical Process, Time Warp cannot guarantee that a set of R-schedulable events can be processed without violating deadlines for any finite R. There has been a rapidly expanding family of Time Warp algorithms focused on constraining the number of false events discussed next.

7.4 CLASSIFICATION OF OPTIMISTIC SIMULATION TECHNIQUES

Another contribution of this section is to classify these algorithms as shown in Figures 7.1, 7.2, 7.3 and Table 7.1. Each new modification to the Time Warp mechanism attempts to improve performance by reducing the expected number of rollbacks. Partitioning methods attempt to divide tasks into logical processes such that the inter-communication is minimized. Also included under partitioning are methods that dynamically move Logical Processes from one processor to another in order to minimize load and/or inter-Logical Process traffic. Delay methods attempt to introduce a minimal amount of wait into Logical Processes such that the increased synchronization and reduced number of rollbacks more than compensates for the added delay. Many of the delay algorithms use some type of windowing method to bound the difference between the fastest and slowest processes.

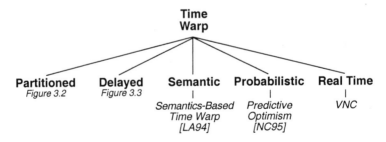

Figure 7.1. Time Warp Family of Algorithms

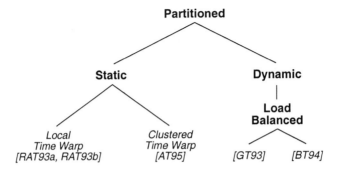

Figure 7.2. Partitioned Algorithms

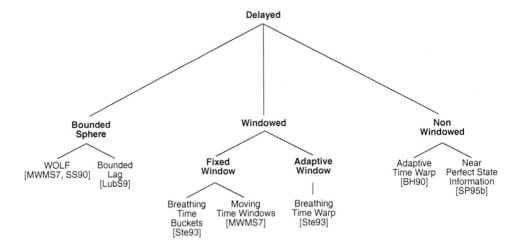

Figure 7.3 Delaying Algorithms

Table 7.1 Time Warp Family of Algorithms.

Class	Sub Class	Sub Class	Description	Example
Probabilistic			Predict msg arrival time.	Predictive Optimism ((Leong and Agrawal, 1994))
Semantic			Contents used to reduce rollback.	Semantics Based Time Warp ((Leong and Agrawal, 1994))
Partitioned			Inter-LP comm minimized.	
	Dynamic		LPs change mode dynamically. LPs migrate across hosts.	
		Load Balanced		((Glazer and Tropper, 1993), and (Boukerche and Tropper, 1994))
	Static		LPs cannot change mode while executing.	Clustered Time Warp ((Avril and Tropper, 1995)) Local Time Warp ((Rajaei et al., 1993a, Rajaei et al., 1993b))
Delayed			Delays reduce rollback.	
	Windowed		Windows reduce rollback.	
		Adaptive Window	Windows adapt to reduce rollback.	Breathing Time Warp ((Steinman, 1993))
		Fixed Window	Window does not adapt.	Breathing Time Buckets ((Steinman, 1993)) Moving Time Windows ((Madisetti et al., 1987))
	Bounded		Based on earliest time inter-LP effects occur.	Bounded Lag ((Lubachevsky, 1989))
	Sphere			WOLF
	Non-Windowed		Non-Window method to reduce rollback.	((Madisetti et al., 1987, Sokol and Stucky, 1990)) Adaptive Time Warp ((Ball and Hoyt, 1990)) Near Perfect State Information ((Srinivisan and Paul F. Reynolds, 1995b))

$$\alpha(i) = \begin{array}{c} \min \\ j \in S \downarrow (i,B) \wedge j \neq i \end{array} \{d(j,i) + \min\{T(j), d(i,j) + T(i)\}\} \qquad (7.1)$$

The bounded sphere class of delay mechanisms attempts to calculate the maximum number of nodes that may need to be rolled back because they have processed messages out of order. For example, $S\downarrow(i, B)$ in (Lubachevsky et al., 1989) is the set of nodes affected by incoming messages from node i in time B, while $S\uparrow(i, B)$ is the set of nodes affected by outgoing messages from node i in time B. The downward pointing arrow in $S\downarrow(i, B)$ indicates incoming messages, while the upward pointing arrow in $S\uparrow(i, B)$ indicates outgoing messages.

Another approach to reducing rollback is to use all available semantic information within messages. For example, commutative sets of messages are messages that may be processed out-of-order yet they produce the same result. Finally, probabilistic methods attempt to predict certain characteristics of the optimistic simulation, usually based on its immediate past history, and take action to reduce rollback based on the predicted characteristic. It is insightful to review a few of these algorithms because they not only trace the development of Time Warp based algorithms but also because they illustrate the "state of the art" in preventing rollback, attempts at improving performance by constraining lookahead, partitioning of Logical Processes into sequential and parallel environments, and the use of semantic information. All of these techniques and more may be applied in the Active Virtual Network Management Prediction algorithm.

The Bounded Lag algorithm (Lubachevsky, 1989) for constraining rollback explicitly calculates, for each Logical Process, the earliest time that an event from another Logical Process may affect the current Logical Process's future. This calculation is done by first determining the $(S\!\downarrow\!(i, B))$, which is the set of nodes that a message may reach in time B. This depends on the minimum propagation delay of a message in simulation time from node i to node j, which is $d(i,j)$. Once $S\!\downarrow\!(i, B)$ is known, the earliest time that node i can be affected, $\alpha(i)$, is shown in Equation 7.1, where $T(i)$ is the minimum message receive time in node i's message receive queue. After processing all messages up to time $\alpha(i)$, all Logical Processes must synchronize.

The Bounded Lag algorithm is conservative because it synchronizes Logical Processes so that no message arrives out of order. The problem is that a minimum $d(i,j)$ must be known and specified before the simulation begins. A large $d(i,j)$ can negate any potential parallelism, because a large $d(i,j)$ implies a large $\alpha(i)$, which implies a longer time period between synchronizations. A filtered rollback extension to Bounded Lag is described in (Lubachevsky et al., 1989). Filtered Rollback allows $d(i,j)$ to be made arbitrarily small, which may possibly generate out of order messages. Thus the basic rollback mechanism described in (Jefferson and Sowizral, 1982) is required.

A thorough understanding of rollbacks and their containment is essential for Active Virtual Network Management Prediction. In (Lubachevsky et al., 1989), rollback cascades are analyzed under the assumption that the Filtered Rollback mechanism is used. Rollback activity is viewed as a tree; a single rollback may cause one or more rollbacks that branch out indefinitely. The analysis is based on a "survival number" of rollback tree branches. The survival number is the difference between the minimum propagation delay $d(j,i)$ and the delay in simulated time for an event at node i to affect the history at node, j $t(i,j)$. Each generation of a rollback caused by an immediately preceding node's rollback adds a positive or negative survival number. These rollbacks can be thought of as a tree whose leaves are rollbacks that have "died out." It is shown that it is possible to calculate upper bounds, namely, infinite or finite number of nodes in the rollback tree.

A probabilistic method is described in (Noble and Chamberlain, 1995). The concept in (Noble and Chamberlain, 1995) is that optimistic simulation mechanisms are making implicit predictions as to when the next message will arrive. A purely optimistic system assumes that if no message has arrived, then no message **will** arrive and computation continues. However, the immediate history of the simulation can be used to attempt to predict when the next message will arrive. This information can be used either for partitioning the location of the Logical Processes on processors or for delaying computation when a message is expected to arrive.

In (McAffer, 1990), a foundation is laid for unifying conservative and optimistic distributed simulation. Risk and aggressiveness are parameters that are explicitly set by the simulation user. Aggressiveness is the parameter controlling the amount of non-causality allowed in order to gain parallelism, and risk is the passing of such results through the simulation system. Both aggressiveness and risk are controlled via a windowing mechanism similar to the sliding lookahead window of the Active Virtual Network Management Prediction algorithm.

A unified framework for conservative and optimistic simulation called ADAPT is described in (Jha and Bagrodia, 1994). ADAPT allows the execution of a "sub-model" to dynamically change from a conservative to an optimistic simulation approach. This is accomplished by uniting conservative and optimistic methods with the same Global Control Mechanism. The mechanism in (Jha and Bagrodia, 1994) has introduced a useful degree of flexibility and described the mechanics for dynamically changing simulation approaches; (Jha and Bagrodia, 1994) does not quantify or discuss the optimal parameter settings for each approach.

A hierarchical method of partitioning Logical Processes is described in (Rajaei et al., 19993a, Rajaei et al., 19993b). The salient feature of this algorithm is to partition Logical Processes into clusters. The Logical Processes operate as in Time Warp. The individual clusters interact with each other in a manner similar to Logical Processes.

The CTW is described in (Avril and Tropper, 1995). The CTW mechanism was developed concurrently but independently of Active Virtual Network Management Prediction. This approach uses Time Warp between clusters of Logical Processes residing on different processors and a sequential algorithm within clusters. This is in some ways similar to the SLogical Process described later in Active Virtual Network Management Prediction. Since the partitioning of the simulation system into clusters is a salient feature of this algorithm, CTW has been categorized as a partitioned algorithm in Figure 7.2. One of the contributions of (Avril and Tropper, 1995) in CTW is an attempt to efficiently control a cluster of Logical Processes on a processor by means of the CE. The CE allows the Logical Processes to behave as individual Logical Processes as in the basic time warp algorithm or as a single collective Logical Process. The algorithm is an optimization method for the Active Virtual Network Management Prediction SLogical Processes.

Semantics Based Time Warp is described in (Leong and Agrawal, 1994). In this algorithm, the Logical Processes are viewed as abstract data type specifications. Messages sent to a Logical Process are viewed as function call arguments and messages received from Logical Processes are viewed as function return values. This allows data type properties such as commutativity to be used to reduce rollback. For example, if commutative messages arrive out-of-order, there is no need for a rollback since the results will be the same.

Another means of reducing rollback, in this case by decreasing the aggressiveness of Time Warp, is given in (Ball and Hoyt, 1990). This scheme involves voluntarily suspending a processor whose rollback rate is too frequent because it is out-pacing its neighbors. Active Virtual Network Management Prediction uses a fixed sliding window to control the rate of forward emulation progress; however, a mechanism based on those just mentioned could be investigated.

The NPSI Adaptive Synchronization Algorithms for Parallel Discrete Event Synchronization are discussed in (Srinivisian and Paul F. Reynolds, 1995a) and (Srinivisian and Paul F. Reynolds, 1995b). The adaptive algorithms use feedback from the simulation itself in order to adapt. Some of the deeper implications of these types of systems are discussed in Appendix 8. The NPSI system requires an overlay system to return feedback information to the Logical Processes. The NPSI Adaptive Synchronization Algorithm examines the system state (or an approximation of the state), calculates an error potential for future error, and then translates the error potential into a value that controls the amount of optimism.

Breathing Time Buckets described in (Steinman, 1992) is one of the simplest fixed window techniques. If there exists a minimum time interval between each event and the earliest event generated by that event (T), then the system runs in time cycles of duration T. All Logical Processes synchronize after each cycle. The problem with this approach is that T must exist and must be known ahead of time. Also, T should be large enough to allow a reasonable amount of parallelism, but not so large as to lose fidelity of the system results.

Breathing Time Warp (Steinman, 1993) attempts to overcome the problems with Breathing Time Buckets and Time Warp by combining the two mechanisms. The simulation mechanism operates in cycles that alternate between a Time Warp phase and a Breathing Time Buckets phase. The reasoning for this mechanism is that messages close to GVT are less likely to cause a rollback, while messages with time-stamps far from GVT are more likely to cause rollback. Breathing Time Warp also introduces the *event horizon*, that is the earliest time of the next new event generated in the current cycle. A user-controlled parameter controls the number of messages that are allowed to be processed beyond GVT. Once this number of messages is generated in the Time Warp phase, the system switches to the Breathing Time Buckets phase. This phase continues to process messages, but does not send any new messages. Once the event horizon is crossed, processing switches back to the Time Warp phase. One can picture the system taking in a breath during the Time Warp phase and exhaling during the Breathing Time Buckets phase.

An attempt to reduce roll-backs is presented in an algorithm called WOLF (Mandisetti et al., 1987, Sokol and Stucky, 1990). This method attempts to maintain a sphere of influence around each rollback in order to limit its effects.

The Moving Time Window (Sokol et al., 1988, Sokol and Stucky, 1990) simulation algorithm is an interesting alternative to Time Warp. It controls the amount of aggressiveness in the system by means of a moving time window MTW. The trade-off in having no roll-backs in this algorithm is loss of fidelity in the simulation results. This could be considered as another method for implementing the Active Virtual Network Management Prediction algorithm.

An adaptive simulation application of Time Warp is presented in (Tinker and Agra, 1990). The idea presented in this paper is to use Time Warp to change the input parameters of a running simulation without having to restart the entire simulation. Also, it is suggested that events external to the simulation can be injected even after that event has been simulated.

Hybrid simulation and real system component models are discussed in (Bagrodia and Shen, 1991). The focus in (Bagrodia and Shen, 1991) is on PIPS Components of a performance specification for a distributed system that are implemented while the remainder of the system is simulated. More components are implemented and tested with the simulated system in an iterative manner until the entire distributed system is implemented. The PIPS system described in (Bagrodia and Shen, 1991) discusses using

MAY or Maisie as a tool to accomplish the task, but does not explicitly discuss Time Warp.

7.5 REAL-TIME CONSTRAINTS IN OPTIMISTIC SIMULATION

The work in (Ghosh et al., 1993) provides some results relevant to Active Virtual Network Management Prediction. It is theorized that if a set of events is R-schedulable in a conservative simulation, and $R \geq \rho + c\ t + \sigma$ where ρ is the time to restore an state, c is the number of Processes, t is the time the simulation has been running, and σ is the real time required to save an state, then the set of events can run to completion without missing any deadline by an NFT Time Warp strategy with lazy cancellation. NFT Time Warp assumes that if an incorrect computation produces an incorrect event $(E_{i,T})$, then it must be the case that the correct computation also produces an event $(E_{i,T})$ with the same timestamp[1]. This result shows that conditions exist in a Time Warp algorithm that guarantee events are able to meet a given deadline. This is encouraging for the Active Virtual Network Management Prediction algorithm since clearly events must be completed before real-time reaches the predicted time of the event for the cached results to be useful in Active Virtual Network Management Prediction. Finally, this author has not been the only one to consider the use of Time Warp to speed up a real-time process. In Tennenhouse and Bose, 1995), the idea of temporal decoupling is applied to a signal processing environment. Differences in granularity of the rate of execution are utilized to cache results before they are needed and to allocate resources more effectively.

This section has shown the results of research into improving Time Warp, especially in reducing rollback, as well as the limited results in applying Time Warp to real time systems. Improvements to Time Warp and the application to real time systems are both directly applicable to Active Virtual Network Management Prediction. Now consider the Active Virtual Network Management Prediction Algorithm in more detail.

7.6 PSEUDOCODE SPECIFICATION FOR AVNMP

The Active Virtual Network Management Prediction algorithm requires both Driving Processes and Logical Processes. Driving Processes predict events and inject virtual messages into the system. Logical Processes react to both real and virtual messages. The Active Virtual Network Management Prediction Algorithm for a driving process is shown in Figure 7.4. The operation of the driving process and the logical process repeat indefinitely. If the Driving Process has not exceeded its lookahead time, a new value Δ time units into the future is computed by the function $C(t)$ and the result is assigned to the message (M) and sent. The receive time, which is the time at which this message value is to be valid, is assigned to (M).

repeat
    if $GVT \leq t + \Lambda$
        then   /* not yet reached lookahead */
            $M.val \leftarrow C(LVT + \Delta)$ /* compute next message
value */
            $M.rt \leftarrow LVT + \Delta$ /* set packet receive time */
            Send(M)
End pseudo-code.

Figure 7.4. AVNMP Driving Process Algorithm.

The Active Virtual Network Management Prediction Algorithm for a Logical Process is specified in Figure 7.5. Note that inf is infimum. The next message from the Receive Queue is checked to determine whether the message is real. If the message is real, the next line in the pseudo-code retrieves the state that was saved closest to the receive time of the message and checks whether the values of the saved state are within tolerance. If the tolerance is exceeded, the process rolls back. Also, if the message is received in the past relative to this process's Local Virtual Time (LVT), the process rolls back as shown. The pre-computed and cached value in the State Queue is committed. Committing a value is an irreversible action because it cannot be rolled back once committed. If the process's Local Virtual Time has not exceeded its time as determined, then the virtual message is processed. The function $C_i(M, LVT)$ represents the computation of the new state. The function $C_i(M, LVT)$ returns the state value for this Logical Process and updates the LVT to the time at which that value is valid. The function $C_2(M, LVT)$ represents the computation of a new message value. The appendix to this chapter takes another look at the algorithm and begins to tie the algorithm to the code provided on the CD included with this book.

$LVT \leftarrow 0$
repeat
        $M \leftarrow \inf M.tr \in QR$ /* retreive message with lowest receive time */
        $CS(t).val \leftarrow C_1(M,t)$/* compute based on new message and update current state */
        if $(M.rt \le t)$ and $(|SQ(t).val - \Theta| > CS(t).val)$
            then Rollback()   /* rollback if real message and out-of-tolerance */
        if $M.rt < LVT$ then Rollback()/* rollback if virtual message and out-of-order */
        if $M.rt \le t$ then Commit($SQ : SQ.t \approx M.rt$)
        if $LVT + \Lambda \le GVT$ then   /* not looking far enough ahead yet */
            $SQ.val \leftarrow C_1(M,LVT)$/* update the state queue with the predicted state */
            $SQ.t \leftarrow LVT$/* record the time of the predicted event */
            $M.val \leftarrow C_2(M,LVT)$ /* generate any new messages based on previous input message */
            $M.rt \leftarrow LVT$ /* set message receive time */
            $QS \leftarrow M$ /* save copy in send queue */
            Send(M)
End pseudo-code.

Figure 7.5. AVNMP Logical Process Algorithm.

Exercises

1. Give a concise definition that compares and contrasts conservative and optimistic simulation techniques.
2. What is the primary problem with the Bounded Lag algorithm?
3. What is the fundamental relationship between conservative synchronization techniques and the ability to take maximum advantage of parallelism?
4. Is GVT necessary to the AVNMP algorithm? Could LVT be used instead?
5. Define super-criticality and give a specific example that demonstrates it.
6. In what ways does an out-of-tolerance rollback differ from an out-of-order rollback?

APPENDIX: AVNMP IMPLEMENTATION

This section discusses enhancing an existing Physical Process (PP)with AVNMP. The web-based tutorial in the CD included with this book provides a step-by-step explanation of how to enhance an application with AVNMP. This appendix provides a more detailed look at the internals of the AVNMP Driving and Logical Processes required in order to perform the enhancement. Notation for Communicating Sequential Processes (CSP) (Hoare, 1981) will serve as an intermediate description before looking at the details of the Java code. In CSP "X?Y" indicates process X will wait until a valid message is received into Y, and "X!Y" indicates X sends message Y. A guard statement is represented by "X→Y," which indicates that condition X must be satisfied in order for Y to be executed. Concurrent operation is indicated by "X $\mid$ $\mid$ Y," which means that X operates in parallel with Y. A "*" preceding a statement indicates that the statement is repeated indefinitely. An alternative command, represented by "X $\mid$ $\mid$ Y," indicates that either X or Y may be executed assuming any guards (conditions) that they may have are satisfied. If X and Y can both be executed, then only one is randomly chosen to execute. A familiar example used to illustrate CSP is shown in Algorithm 7.A.1. This is the bounded buffer problem in which a finite size buffer requests more items from a consumer only when the buffer will not run out of capacity.

Assume a working PP abstracted in Algorithm 7.A.2 where S and D represent the source and destination of real and virtual messages. Algorithm 7.A.3 shows the PP converted to a AVNMP LP operating with a monotonically increasing LVT. Note that the actual AVNMP Class function names are used; however, all the function arguments are not shown in order to simplify the explanation. Each function is described in more detail later. The input messages are queued in the Receive Queue as shown in Algorithm 7.A.3 by recvm(). In non-rollback operation the function getnextvm() returns the next valid message from the Receive Queue to be processed by the PP. When the PP has a message to be sent, the message is place in the State Queue by sendvm(). While a message is flowing through the process, the process saves its state periodically. Normal operation of the AVNMP as just described may be interrupted by a rollback. If recvm() returns a non-zero value, then either an out-of-order or out-of-tolerance message has been received. In order to perform the rollback, getstate() is called to return the proper state to which the process must rollback. It is the application's responsibility to ensure that the data returned from getstate() properly restores the process state. Anti-messages are sent by repeatedly calling rbq() until rbq() returns a null value. With each call of rbq(), an anti-message is returned which is sent to the destination of the original message.

7.A.1. AVNMP Class Implementation

Figure 7.A.4 lists a selection of the main classes and their primary purpose in the AVNMP system. A complete list of the classes and their descriptions can be found on the CD in README.html. The classes are the primary classes for understanding the operation of the AVNMP system.

```
X::
buffer:(0..9) portion;
in,out:integer; in := 0; out := 0;
*[in < out + 10; producer?buffer(in mod 10) →
 in := in + 1;
|| out < in; consumer?more() →
 consumer!buffer(out mod 10); out := out + 1;
]
```

Figure 7.A.1. A CSP Example

```
PP::
*[S?input;
  output := process(input);
  D!output]
```

Figure 7.A.2. A Physical Process

```
PP::
*[S?input;
[recvm(input)!=0 → getstate();
*[rbq()!=NULL → S!AvnmpDriverRb;D!rbq()]||
[recvm(input)==0 →
savestate();
input := getnextvm();
output := process(input);
sendvm(output);
D!output]
]
]
```

Figure 7.A.3. The Logical Process

avnmp.java.lp.AvnmpRecQueue Receive a message, determine whether virtual or real, rollback

avnmp.java.lp.AvnmpSndQueue Send a virtual message, save a copy

avnmp.java.lp.AvnmpQueue All queue related functions

avnmp.java.lp.AvnmpLP Roll back to given time

avnmp.java.lp.AvnmpStateQueue Save previous state

avnmp.java.lp.AvnmpTime Local virtual time maintenance functions

avnmp.java.lp.AvnmpPacket The virtual message

avnmp.java.dp.Driver The driving process

avnmp.java.pp.PP The physical process

avnmp.java.pp.PayLoad The real message

Figure 7.A.4. AVNMP Class Files

Predict() $\to$ output $\to$ getvm()

Figure 7.A.5. The Driving Process

input $\to$ process $\to$ output

Figure 7.A.6. The Logical Process

7.A.2 AVNMP Logical Process Implementation

This class implements the AVNMP logical process. The general idea is to have a working process modified in Figure 7.A.6. Figure 7.A.7 shows the "normal" operation, while Figure 7.A.8 shows the operation of the process when a rollback occurs.

$$\text{input} \to \text{getvm}(); \text{getnext}() \to$$
$$\begin{cases} \text{process} \to \text{sendvm}() \to \text{output} \\ \text{savestate}() \end{cases}$$

Figure 7.A.7. AVNMP Normal Operation

$$\text{if}(\text{getvm}() \neq 0)\text{getstate}() \to \text{process}() \to \text{rbq} \to \text{output}$$

Figure 7.A.8. AVNMP Rollback Operation

Notes

[1]This simplification makes the analysis in (Ghosh et al., 1993) tractable. This assumption also greatly simplifies the analysis of Active Virtual Network Management Prediction. The Active Virtual Network Management Prediction algorithm is simplified because the state verification component of Active Virtual Network Management Prediction requires that saved states be compared with the real-time state of the process. This is done easily under the assumption that the T (timestamp) values of the two events $E_{i,Tv}$ and $E_{i,Tr}$ are the same.

8

ALGORITHM ANALYSIS

The purpose of this section is to analyze the performance of the Active Virtual Network Management Prediction Algorithm. As discussed in detail in previous chapters, current network management is centralized, as shown in Figure 8.1. On the other hand, the Active Virtual Network Management Prediction Algorithm distributes management. Figure 8.2 shows an active network testbed consisting of three active nodes. The active nodes are labeled AN-1, AN-4, and AN-5, and the links are labeled L-1, L-2, L-3, and L-4. One of the goals of this section is to investigate the benefits of the new active network based distributed management model. The characteristics of the Active Virtual Network Management Prediction Algorithm analyzed in this section are speedup, lookahead, accuracy, and overhead. Speedup is the ratio of the time required to perform an operation without the Active Virtual Network Management Prediction Algorithm to the time required with the Active Virtual Network Management Prediction Algorithm. Lookahead is the distance into the future that the system can predict events. Accuracy is related to the rate of convergence between the predicted and actual values. Bandwidth overhead is the ratio of the amount of additional bandwidth required by the Active Virtual Network Management Prediction Algorithm system to the amount of bandwidth required without the Active Virtual Network Management Prediction Algorithm system, and processing overhead is the reduction in network capacity due to active packet execution.

Because the Logical Processes of the Active Virtual Network Management Prediction Algorithm system are asynchronous, they can take maximum advantage of parallelism. However, messages among processes may arrive at a destination process out-of-order as illustrated in Figure 5.2. As shown in Figure 8.2, a virtual network representing the actual network can be viewed as overlaying the actual network for analytical purposes.

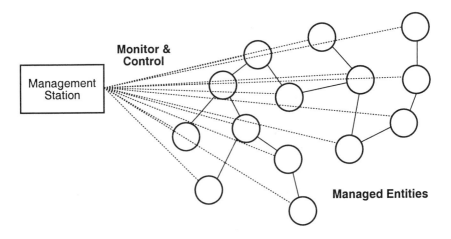

Figure 8.1. Centralized Network Management Model.

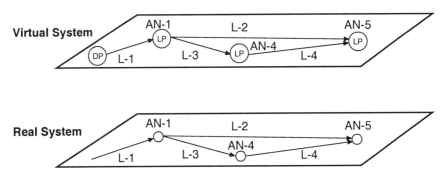

Figure 8.2. AVNMP as a Virtual Overlay for Network Management.

Virtual messages may not arrive at a logical process in the order of Receive Time for several reasons. The first reason is that in an optimistic parallel model, virtual messages are executed as soon as they arrive at a logical process. Thus, in an optimistic simulation of a complex network, virtual messages do not block or delay to enforce causality. This leads to a possibility of messages arriving out-of-order even if the virtual message links have no transmission delay. Petri-Net theory is used to analyze this type of out-of-order message arrival. Petri-Nets are commonly used for synchronization analysis. In Petri-Nets, "places," usually shown as circles, represent entities such as producers, consumers, or buffers, and "transitions," shown as squares, allow "tokens," shown as dots, to move from one place to another. In this analysis, tokens represent the Active Virtual Network Management Prediction Algorithm messages and Petri-Net places represent the Active

Virtual Network Management Prediction Algorithm Logical Processes. Characteristics of Petri-Nets are used to determine the likelihood of out-of-order messages.

Another source of out-of-order virtual message arrival at a logical process is due to congestion or queuing delay. The actual messages in Figure 8.2 can cause the virtual messages along a particular link to arrive later than virtual messages arriving along another link to the same logical process. However, the Active Virtual Network Management Prediction Algorithm can predict that the congestion and thus the late virtual message arrival are likely to occur. The accuracy of this prediction depends in part upon the acceptable tolerance setting of the prediction. The relationship of the tolerance to prediction accuracy and late virtual message arrival likelihood are discussed later in this chapter. If a Logical Process predicts congestion along an input link, then the Logical Process delays itself until some virtual message arrives along that link, thus avoiding a possible rollback. The likelihood of the occurrence of out-of-order messages and out-of-tolerance messages is required by an equation that is developed in this chapter to describe the speedup of the Active Virtual Network Management Prediction Algorithm. After analyzing the speedup and lookahead, the prediction accuracy and overhead are analyzed. This chapter considers enhancements and optimizations such as implementing multiple future events, eliminating the calculation, and elimination of real messages when they are not required.

Performance analysis of the Active Virtual Network Management Prediction algorithm must take into account accuracy as well as distance into the future that predictions are made. An inaccurate prediction can result in committed resources that are never used and thus wasted, or in not committing enough resources when needed, thus causing a delay. Unused resource allocation must be minimized. Active Virtual Network Management Prediction does not require permanent over-allocation of resources; however, the Active Virtual Network Management Prediction algorithm may make a false prediction that *temporarily* establishes resources that may never be used. An Active Virtual Network Management Prediction system whose tolerances are reduced in order to produce more accurate results will have fewer unused allocated resources; however, the tradeoff is a reduction in speedup.

$$U_{AVNMP} = \eta\,\Phi_s - \alpha\,\Phi_w - \beta\,\Phi_b \qquad (8.1)$$

Equation (8.1) quantifies the advantage of using Active Virtual Network Management Prediction where η is the expected speedup using Active Virtual Network Management Prediction over a non-Active Virtual Network Management Prediction system, Φ_s is the marginal utility function of the configuration speed, and α is the expected quantity of wasted resources other than overhead, and Φ_w is the marginal utility function of the allocated but unused resource. An example of a resource that may be temporarily wasted due to prediction error is a Virtual Circuit in a mobile wireless network that may be established temporarily and never used. The expected overhead is represented by β and Φ_b is the marginal utility function of bandwidth and processing.

The marginal utility functions Φ_s, Φ_w and Φ_b are subjective functions that describe the value of a particular service to the user. The functions Φ_s, Φ_w and Φ_b may be determined by monetary considerations and user perceptions. The following sections develop propositions that describe the behavior of the Active Virtual Network Management Prediction algorithm and from these propositions equations for η, α and β are defined.

8.1 PETRI-NET ANALYSIS FOR THE AVNMP ALGORITHM

In this section the probability of message arrival at a Logical Process is determined, the expected proportion of messages ($E[X]$) and the probability of rollback due to messages (P_{oo}) is analyzed, and a new and simpler approach to analyzing Time-Warp based algorithms in general and the Active Virtual Network Management Prediction Algorithm in particular is developed. The contribution is unique because most current optimistic analysis has been explicitly time-based, yielding limited results except for very specific cases. The approach is topological; timing is implicit rather than explicit. A C/E is used in this analysis because it is the simplest form of a Petri-Net that is ideal for studying the Active Virtual Network Management Prediction Algorithm synchronization behavior.

A C/E network consists of condition and transition elements that contain tokens. Tokens reside in condition elements. When all condition elements leading to a transition element contain a token, several changes take place in the network. First, the tokens are removed from the conditions that triggered the event, the event occurs, and finally tokens are placed in all condition outputs from the transition that was triggered. Multiple tokens in a condition and the uniqueness of the tokens is irrelevant in a C/E Net. In this analysis, tokens represent virtual messages, conditions represent processes, and transitions represent interconnections. The notation from (Reisig, 1985) is used: $\Sigma = (B,E;F,C)$ is a C/E Net where B is the set of conditions, E is the set of transitions, and $F \subseteq (B \times E) \cup (E \times B)$ where $\cup$ is union and $\times$ is the cross product of all conditions and transitions. A marking is the set of conditions containing tokens at any given time during C/E operation and C is the set of all possible sets of markings of Σ. The input conditions to a transition are written as "pre$-e$" and the output conditions are written as "post$-e$." Let $c \subseteq C$, then a transition $e \in E$ is triggered when pre$-e \subseteq (c \subseteq B)$ and post$-e \cap c = \varnothing$. If c is the current set of enabled conditions and after the next transition (e) the new set of enabled conditions is c', then this is represented more compactly as $c[e \, \rangle c'$. C/E networks provide insight into liveness, isomorphism, reachability, a method for determining synchronous behavior, and behavior based on the topology of the Active Virtual Network Management Prediction Algorithm Logical Process communication. Every Finite State Machine has an equivalent C/E Net (Peterson, 1981, p. 42).

Some common terminology and concepts are defined next that are needed for a topological analysis of the Active Virtual Network Management Prediction Algorithm. These terms and concepts are introduced in a brief manner and build upon one another. Their relationship with the Active Virtual Network Management Prediction Algorithm will soon be made clear. The following notation is used: "$\neg$" means "logical not," "$\exists$" means "there exists," "$\forall$" means "for each," "$\wedge$" means "logical and,", "$\vee$" means "logical or," "$\in$" means that an element is a member of a set, "$\equiv$" means "defined as," and "$\rightarrow$" defines a mapping or function. Also, $a \prec b$ indicates an ordering between two elements, *a* and *b*, such that *a* precedes *b* in some relation. "$\Rightarrow$" means "logical implication" and "$\leftrightarrow$" means "logical equivalence."

A region of a particular similarity relation ($\cdot$) of $B \subseteq A$ means that $\forall a,b \in B : a \cdot b$ and $\forall a \in A : a \notin B \Rightarrow \exists b \in B : \neg (a \cdot b)$. This means that the relation is "full" on B and B is a maximal subset on which the relation is full. In other words, a graph of the relation ($\cdot$) would show B as the largest fully connected subset of nodes in A.

Let "$\underline{li}$" represent a such that a $\underline{li}$ $b \leftrightarrow (a \prec b) \vee (b \prec a) \vee (a \equiv b)$. Let "$\underline{co}$" represent a concurrent ordering a $\underline{co}$ b $\leftrightarrow \neg (a$ $\underline{li}$ b$) \vee (a \equiv b)$. Figure 8.3 illustrates a region of $\underline{co}$ that contains $\{a, c\}$ and of $\underline{li}$ that contains $\{a, b, d\}$ where $\{a, b, c, d\}$ represents Logical Processes and the relation is "sends a message to." Trivially, if every process in the Active Virtual Network Management Prediction Algorithm system is a region of $\underline{li}$ then regardless of how many driving processes there are, no synchronization is necessary since there exist no processes. If no synchronization is needed, then virtual messages cannot arrive out-of-order; thus no rollback will occur.

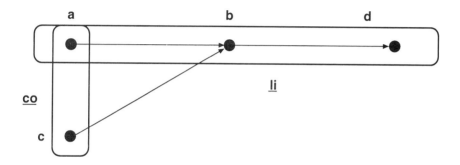

Figure 8.3. Demonstration of $\underline{li}$ and $\underline{co}$.

Let D be the set of driving processes and R be the set of the remaining processes in the Active Virtual Network Management Prediction Algorithm system. Then $D \prec R \leftrightarrow \forall d \in D \ \forall r \in R : (d \prec r) \vee (d \ \underline{co} \ r)$. In order for the virtual messages that originate from D to be used, $D \prec R$ where R are the remaining non-driving processes. This is again assumed to be "sends a message to."

In the remaining definitions, let A, B, and C be arbitrary sets where $B \subseteq A$ used for defining additional operators. Let $B \preceq C \equiv \forall b \in B \ \forall c \in C: b \prec c \vee b \ \underline{co} \ c$. Let $B^- \equiv \{ a \in A \mid \{a\} \preceq B \}$ and $B^+ \equiv \{ a \in A \mid B \preceq \{a\} \}$ where | means "such that." Also, let $[\bar{B}] \equiv \{b \in B \mid \forall b' \in B: (b \ \underline{co} \ b') \vee (b \prec b') \}$ and $\underline{B} \equiv \{ b \in B \mid \forall b' \in B: (b \ \underline{co} \ b') \vee (b' \prec b) \}$. This is illustrated in Figure 8.4, where all nodes are in the set A and B is the set of nodes that lie within the circle. B^- is the set $\{a,b,c,d,f\}$ and $[\bar{B}]$ is the set $\{b\}$.

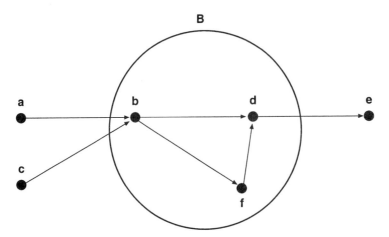

Figure 8.4. Illustration of B⁻ and [B̄].

An occurrence network (*K*) is a network that is related to the operation of a particular network (Σ). The occurrence network (*K*) begins as an empty C/E network; conditions and events are added to *K* as Σ operates. *K* represents a particular sample of operation of Σ. There can be multiple events in Σ that are capable of firing, but only one event is chosen to fire; thus it is possible that a particular Σ will not always generate the same occurrence net (*K*) each time it operates. Note that *K* has some special properties. The condition elements of *K* have one and only one transition, because only one token in Σ may fire from a given condition. Also, *K* is cycle free because *K* represents the operation of Σ.

A few more definitions are required before the relation described above between *K* and Σ can be formally defined. This relationship is called a Petri-Net process. Once the Petri-Net process is defined, a measure for the "out-of-orderness" of messages can be developed based on synchronic distance. A *line* is a subset that is a region of li and a *cut* is a subset that is a region of co. A slice ("sl") is a cut of an occurrence network (*K*) containing condition elements, and sl (*K*) is the set of *all* slices of *K*. The of co shown in Figure 8.3 illustrates a cut where nodes represent conditions and the relation defines an event from one condition to another in a C/E Network.

A formal definition of the relation between an occurrence net and a C/E net is given by a Petri-Net process. A Petri-Net process (*p*) is defined as a mapping from a network *K* to a C/E Network Σ, *p* : *K* → Σ, such that each slice of *K* is mapped injectively (one-to-one) into a marking and (*p*(pre−*t*) = pre−*p*(*t*)) ∧(*p*(post−*t*) = post−*p*(*t*)). Also note that p^{-1} is used to indicate the inverse mapping of *p*. Think of *K* as a particular sample of the operation of a C/E Network. A C/E Network can generate multiple processes. Another useful characteristic is whether a network is K-dense. A network is K-dense if and only if every sl (*K*) has a non-empty intersection with every region of li in *K*. This means that each intersects every sequential path of operation.

All of the preceding definitions have been leading towards the development of a measure for the "out-of-orderness" of messages that does not rely on explicit time values or distributions. In the following explanation, a measure is developed for the synchronization between events. Consider D_1 and D_2 that are two slices of K and M is a set of events in a C/E Network. $\mu(M, D_1, D_2)$ is defined as $|\ M \cap D^+_2 \cap D^-_1\ | - |\ M \cap D^-_1 \cap D^+_2\ |$. Note that $\mu(M, D_1, D_2) = -\ \mu(M, D_2, D_1)$. Thus $\mu(M, D_1, D_2)$ is a number that defines the number of events between two specific slices of a net.

Let $(p:K \to \Sigma) \in \pi_\Sigma$ where π_Σ is the set of all finite processes of Σ. A term known as "variance" is defined that describes the number of events across all slices of a net (K). The variance of T_Σ is $v(p, T_1, T_2) \equiv \max\{\mu(p^{-1}(T_1), D_1, D_2) - \mu(p^{-1}(T_2), D_1, D_2) \mid D_1, D_2 \in$ $\underline{sl}\ (K)\}$. Also, note that $v(p, T_1, T_2) = v(p, T_2, T_1)$ where and $T_1, T_2 \subseteq T_\Sigma$. This defines a measure of the number of events across all slices of a net (K).

The synchronic distance $(\sigma(T_1, T_2) = \sup\{\ v(p, T_1, T_2) \mid p \in \pi_\Sigma\ \})$ is the supremum of the variance in all finite processes. This defines the measure of "out-of-orderness" across all possible K. By determining the synchronic distance, a measure for the likelihood of rollback in the Active Virtual Network Management Prediction Algorithm can be defined that is dependent on the topology and is **independent of time**. Further details on syn chronic distance and the relation of synchronic distance to other measures of synchrony can be found in (Voss et al, 1987). A more intuitive method for calculating the synchronic distance is to insert a virtual condition into the C/E net. This condition has no meaning or effect on operation. The condition is allowed to hold multiple tokens and begins with enough tokens so that it can emit a token whenever a condition connected to its output transition is ready to fire. The virtual condition has inputs from all members of T_1 and output transitions of all members of T_2. The synchronic distance is the maximum variation in the number of tokens in the virtual condition. The greater the possibility of rollback, the larger the value of $\sigma(T_1, T_2)$. A simple example in Figure 8.5 intuitively illustrates what the synchronic distance means. Using the virtual condition method to calculate the synchronic distance between $\{a, b\}$ and $\{c, d\}$ in the upper C/E Network, the synchronic distance is found to be two. By adding two more conditions and another transition to the C/E network, the synchronic distance of the lower C/E Network shown in Figure 8.5 is one. The larger the value of $\sigma(T_1, T_2)$, the less synchronized the events in sets T_1 and T_2. If these events indicate message transmission, then the less synchronized the events, the greater the likelihood that the messages based on events T_1 and T_2 are out-of-order. This allows the likelihood of message arrival at a Logical Process to be determined based on the inherent synchronization of a system. However, a completely synchronized system does not gain the full potential provided by optimistic parallel synchronization.

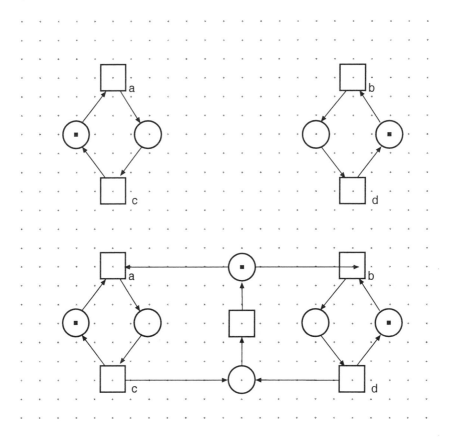

Figure 8.5. Example of Synchronic Distance.

A P/T Network is similar to a C/E network except that a P/T Net allows multiple tokens in a place and multiple tokens may be required to cause a transition to fire. Places are defined by the set S and transitions by the set T. The operation of a network can be described by a matrix. The rows of the matrix represent places and the columns represent transitions. The last column of the matrix represents the current number of tokens in a place. Each element of the matrix contains the number of tokens that either leave (negative integer) or enter (positive integer) a place when the transition fires. When a transition fires, the column corresponding to the transition is added to the last column of the matrix. The last column of the matrix changes as the number of nodes in each place change. The matrix representation of a P/T Network is shown in Matrix 8.2, where $LP_n \in S$, $c_n \in T$ and $w_{i,j}$ is the weight or number of tokens required by link j to fire or the number of tokens generated by place i. Note that LP_n and c_n bordering Matrix 8.2 indicate labels for rows and columns. Note also that there exists a duality between places and transitions such that places and transitions can be interchanged (Peterson, 1981, p. 13). P/T networks can be extended from the state representation of C/E networks to examine problems involving quantities of elements in a system, such as producer/consumer problems. The places in this analysis are analogous to Logical Processes because they produce and consume both real and virtual messages. Transitions in this analysis are analogous to connections between Logical Processes, and tokens to messages. The weight, or number of tokens, is $-w_{i,j}$ for outgoing tokens and $w_{i,j}$ for incoming tokens. The current marking, or expected value of the number of tokens held in each place, is given in column vector m_N. A transition to the next state is determined by $m_{N+1} = m_N + c_i$ where c_i is the column vector of the transition that fired and N is the current matrix index.

$$M_N = \begin{array}{c} \\ LP_1 \\ LP_2 \\ M_N = \ \ LP_3 \\ \vdots \end{array} \begin{pmatrix} c_1 & c_2 & c_3 & \cdots & m_N \\ w_{1,1} & w_{1,2} & w_{1,3} & \cdots & n_1 \\ w_{2,1} & w_{2,2} & w_{2,3} & \cdots & n_2 \\ w_{3,1} & w_{3,2} & w_{3,3} & \cdots & n_3 \\ \vdots & \vdots & \vdots & \vdots & \vdots \end{pmatrix} \tag{8.2}$$

A global synchronic distance value is shown in Equation 8.3 where T consists of the set of all transitions. The global synchronic distance is used to define a normalized measure. The global measure is the maximum in a P/T network and $\sigma_n(I_1,I_2) \in [0,1]$ is a normalized value shown in Equation 8.4 where $\{I_n\}$ is a set of all incoming transitions to a particular place. A probability of being within tolerance is defined in vector p shown in Matrix 8.5. Each LP_i along the side of Matrix 8.5 indicates a LP and the $1 - P_{ot}$ along the top of Matrix 8.5 indicates p_i values that are the individual probabilities that the tolerance is not exceeded. The probability of out-of-tolerance rollback is discussed in more detail in Section 4.1. Let (LP_i, c_j) be the transition from LP_i across connection c_j. After each transition of M_N from (LP_i, c_j), the next value of n_i that is the element in the ith row of the last column of M_N is $\sigma_n(I_1,I_2)\, p_i^{\,n_i}$.

$$GSV = \max_{f_1, f_2 \subset T} \{\sigma(f_1, f_2)\} \tag{8.3}$$

$$\sigma_n(I_1, I_2) = 1.0 - \frac{\sigma(I_1, I_2)}{GSV} \tag{8.4}$$

$$\vec{p} = \begin{array}{c} \\ LP_1 \\ LP_2 \\ LP_3 \\ \vdots \end{array} \begin{pmatrix} 1 - P_{ot} \\ p_1 \\ p_2 \\ p_3 \\ \vdots \end{pmatrix} \tag{8.5}$$

It is possible for the synchronic distance to be infinite. One way to avoid an infinite synchronic distance is to use weighted synchronic distances. A brief overview of weighted synchronic distances is given in this section. (Andre et al., 1979) introduces capacity Petri-Nets (CPN). Capacity Petri-Nets have place values that hold a multiple number of tokens but with a maximum capacity. A transition cannot fire if it results in a place exceeding its pre-specified capacity. The capacity has an effect upon the synchronic distance. A place between two sets of transitions enforces a synchronic distance equal to the capacity of that place. This is directly apparent because an intuitive method for determining synchronic distance is to add a place with inputs from one set of transitions and outputs to the other set. The synchronic distance is the maximum number of tokens that can appear in the place given all possible firing sequences. In (Goltz and Reisig, 1982) weighted synchronic distances are introduced. Synchronic distance as originally defined can in many instances become infinite even though it is apparent a regular structure exists in the Petri-Net. In (Goltz, 1987) the concept of synchronic distance is introduced along with weighted synchronic distance. (Silva and Colom, 1988) builds on the relationship between synchronic invariants and linear programming. In (Silva and Murata, 1992) measures related to synchronic distances are discussed, namely bounded-fairness. Bounded-fair relations are concerned with the number of times a transition fires before another transition can fire. Marked graphs form a subset of Petri-Nets. The synchronic distance matrix of a marked graph holds the synchronic distances between every vertex in the marked graph. In (Mikami et al., 1993, Tamura and Abe, 1996) necessary and sufficient conditions are given for a matrix to represent a marked graph.

As $p_i^{n_i}$ approaches zero, the likelihood of an out-of-tolerance induced rollback increases. As $\sigma_n(I_1,I_2)\, p_i^{n_i}$ becomes very small, the likelihood of a rollback increases either due to a violation of causality or an out-of-tolerance state value. Synchronic distance is a metric and furthermore the $\sigma_n(I_1,I_2)$ value is treated as a probability because it has the axiomatic properties of a probability. The axiomatic properties are that $\sigma_n(I_1,I_2)$ assigns a number greater than or equal to zero to each synchronic value, $\sigma_n(I_1,I_2)$ has the value of one when messages are always in order, and $\sigma_n(A) + \sigma_n(B) = \sigma_n(A \cup B)$, where A and B are mutually exclusive sets of transitions.

A brief example is shown in Figure 8.6. The initial state shown in Figure 8.6 is represented in Matrix 8.6. The Global Synchronic Value of this network is four. The tolerance vector for this example is shown in Vector 8.7. Consider transition a shown in Figure 8.6; it is enabled since tokens are available in all of its inputs. The element in the p column vector shown in Vector 8.7 is taken to the power of the corresponding elements of the column vector $\vec{a}$ in Matrix 8.6 that are greater than zero ($p_i^{n_i}$). This is the probability that all messages passing through transition a arrive within tolerance. All columns of rows of $\vec{a}$ that are greater than zero that have greater than zero values form the input set ($\{I_n\}$) for $\sigma_n(I_1,I_2)$. Since transition a has only one input, $\sigma_n(\{a\})$ is one. When transition a fires, column vector $\vec{a}$ is added to column vector m_0 to generate a new vector m_1. Matrix 8.8 results after transition a fires. Continuing in this manner, Matrix shows the result after transition b fires. Since $\sigma_n(\{b\})$ is one, row LP_4 of m_2 is 0.3.

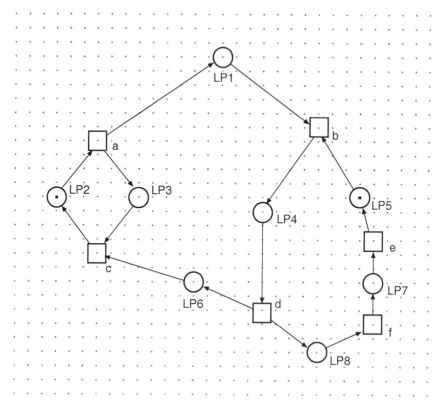

Figure 8.6. Example of P_{oo} Analysis.

$$
M_0 =
\begin{array}{c}
\\
LP_1 \\
LP_2 \\
LP_3 \\
LP_4 \\
LP_5 \\
LP_6 \\
LP_7 \\
LP_8
\end{array}
\begin{array}{c}
\begin{array}{ccccccc}
a & b & c & d & e & f & m_0
\end{array} \\
\left(
\begin{array}{ccccccc}
1 & -1 & 0 & 0 & 1 & 0 & 0 \\
-1 & 0 & 1 & 0 & 0 & 0 & 1 \\
1 & 0 & -1 & 0 & 0 & 0 & 0 \\
0 & 1 & 0 & -1 & 0 & 0 & 0 \\
0 & -1 & 0 & 0 & 1 & 0 & 1 \\
0 & 0 & -1 & 1 & 0 & 0 & 0 \\
0 & 0 & 0 & 0 & -1 & 1 & 0 \\
0 & 0 & 0 & 1 & 0 & -1 & 0
\end{array}
\right)
\end{array}
\qquad (8.6)
$$

$$
\vec{p} =
\begin{array}{c}
 \\
LP_1 \\
LP_2 \\
LP_3 \\
LP_4 \\
LP_5 \\
LP_6 \\
LP_7 \\
LP_8
\end{array}
\begin{array}{c}
1 - P_{ot} \\
\left(\begin{array}{c}
0.7 \\
0.2 \\
0.3 \\
0.4 \\
0.6 \\
0.4 \\
0.2 \\
0.1
\end{array}\right)
\end{array}
\tag{8.7}
$$

$$
M_1 =
\begin{array}{c}
 \\
LP_1 \\
LP_2 \\
LP_3 \\
LP_4 \\
LP_5 \\
LP_6 \\
LP_7 \\
LP_8
\end{array}
\begin{array}{ccccccc}
a & b & c & d & e & f & m_1 \\
\left(\begin{array}{ccccccc}
1 & -1 & 0 & 0 & 1 & 0 & 0.7 \\
-1 & 0 & 1 & 0 & 0 & 0 & 0 \\
1 & 0 & -1 & 0 & 0 & 0 & 0.3 \\
0 & 1 & 0 & -1 & 0 & 0 & 0 \\
0 & -1 & 0 & 0 & 1 & 0 & 1 \\
0 & 0 & -1 & 1 & 0 & 0 & 0 \\
0 & 0 & 0 & 0 & -1 & 1 & 0 \\
0 & 0 & 0 & 1 & 0 & -1 & 0
\end{array}\right)
\end{array}
\tag{8.8}
$$

$$
M_2 =
\begin{array}{c}
 \\
LP_1 \\
LP_2 \\
LP_3 \\
LP_4 \\
LP_5 \\
LP_6 \\
LP_7 \\
LP_8
\end{array}
\begin{array}{ccccccc}
a & b & c & d & e & f & m_2 \\
\left(\begin{array}{ccccccc}
1 & -1 & 0 & 0 & 1 & 0 & 0 \\
-1 & 0 & 1 & 0 & 0 & 0 & 0 \\
1 & 0 & -1 & 0 & 0 & 0 & 0.3 \\
0 & 1 & 0 & -1 & 0 & 0 & 0.3 \\
0 & -1 & 0 & 0 & 1 & 0 & 0 \\
0 & 0 & -1 & 1 & 0 & 0 & 0 \\
0 & 0 & 0 & 0 & -1 & 1 & 0 \\
0 & 0 & 0 & 1 & 0 & -1 & 0
\end{array}\right)
\end{array}
\tag{8.9}
$$

The analysis presented in this section reduces the time and topological complexities characteristic of more explicit time analysis methods to simpler and more insightful matrix manipulations. The method presented is used in the following section to determine the probability of rollback due to messages, $P_{oo} = 1 - \sigma_n(I_1,I_2)$.

Also, the worst case proportion of out-of-order messages (X) is calculated as follows. The ($\sigma(I_1,I_2)$) is a measure of the maximum difference in the rate of firing among transitions. The maximum possible value of $\sigma(I_1,I_2)$ that can occur is the rate of the slowest firing transition in sets I_1,I_2. Equation 8.10 shows the relationship between $E[X]$ and the rate at which transition I fires.

$$E[X] \leq \min_{\{Transition \in I_1,I_2\}} \{rate\ (Transition)\} \qquad (8.10)$$

8.1.1 T-Invariants

An alternative analysis of the likelihood of out-of-order message arrival at a logical process and quantitative synchronization analysis can be derived from invariants in the Petri-Net representation of the Active Virtual Network Management Predication system. T-invariants are transition vectors whose values are the number of times each transition fires in order to obtain the same marking. P-variants are sets of places that always contain the same number of tokens. In (J. Martinez and Silva, 1982) an algorithm is given to determine all the invariants of generalized and capacity Petri-Nets.

Figure 8.7 provides an example of a sample active network not yet enhanced with Active Virtual Network Management Prediction. The active network nodes are illustrated as well as the end-systems and the active packet. A Petri-Net representation of this network is derived as follows. The logical processes are injected into the network and persist at the active nodes to be AVNMP-enhanced as shown in Figure 8.8. The Active Virtual Network Management Prediction system was developed using the Magician (Kulkarni et al., 1998) execution environment; the driving processes, logical processes, and virtual messages are implemented as active packets. The driving processes reside at the edge of the region to be enhanced with AVNMP. Virtual messages now enter the picture.

This analysis considers the number of transition firings as the local virtual time. Thus, the logical processes are transitions. The token represents an update to the local virtual time of the logical process driven by the receive time of a virtual message that has been processed. Thus, in the transition from Figure 8.8 to 8.9, the driving processes become token generators and logical processes become Petri-Net transitions. The active packets that were virtual messages become Petri-Net tokens.

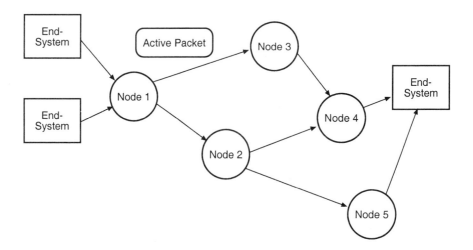

Figure 8.7. Active Network Configuration for T-Invariant Analysis.

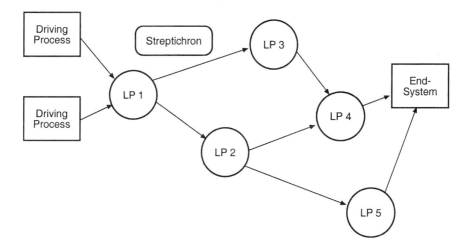

Figure 8.8. Active Network with AVNMP for T-Invariant Analysis.

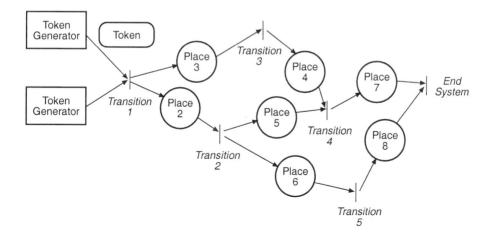

Figure 8.9. Petri-Net Representation of Active Network with AVNMP for T-Invariant Analysis.

A rollback occurs when an incoming virtual message has a less than the logical process's. The receive time of a virtual message is determined by the local virtual time of the sending logical process. It is assumed that the receive time cannot be less than the local virtual time of the sending logical process. Let T_j be the total number of transition firings for logical process j. When a token arrives at logical process j from logical process i, a rollback does not occur as long as $T_i < T_j$. A logical process can receive virtual messages from more than one logical process. Let T_i^* be the set of all inputs to logical process j. Then $\forall T_k \in T_i^*: T_k < T_j$. If N is a matrix form of the Petri-Net as used in the previous section, Matrix 8.6 for example, and x is a vector of transitions, the T-Invariant is computed as shown in Equation 8.11. Based upon the set of x that satisfy 8.11, it is possible to determine whether j will rollback, and if so, how many of the possible invariants cause a rollback.

$$N \cdot x = 0 \qquad\qquad (8.11)$$

8.2 EXPECTED SPEEDUP: η

This section analyzes the primary benefit of Active Virtual Network Management Prediction, namely expected lookahead into the future. This depends on the rate that the system can generate and handle predictions. This rate is referred to as speedup, because when these values were cached and used, they increase the rate at which the system executes. There are many factors which influence speedup including out-of-order message probability, out-of-tolerance state value probability, rate of virtual messages entering the system, task execution time, task partitioning into Logical Processes, rollback overhead, prediction accuracy as a function of the distance into the future which predictions are attempted, and the effects of parallelism and optimistic synchronization.

All of these factors are considered, beginning with a direct analysis using definitions from optimistic simulation.

The definition of Global Virtual Time (GVT) can be applied to determine the relationship among expected task execution time (τ_{task}), the real time at which the state was cached (t_{SQ}), and real time (t). Consider the value (V_v), which is cached at real time t_{SQ} in the SQ resulting from a particular predicted event. For example, refer to Figures 5.16 through 5.20 and notice that state queue values may be repeatedly added and discarded as Active Virtual Network Management Prediction operation proceeds in the presence of rollback. As rollbacks occur, values for a particular predicted event may change, converging to the real value (V_r). For correct operation of Active Virtual Network Management Prediction, V_v should approach V_r as t approaches $GVT(t)$ where $GVT(t)$ is the GVT of the Active Virtual Network Management Prediction system at time t. Explicitly, this is $\forall \varepsilon > 0 \ \exists \delta > 0$ s.t. $|f(t) - f(GVT(t))| < \varepsilon \Rightarrow 0 < |GVT(t) - t| < \delta$ where $f(t) = V_r$ and $f(GVT(t)) = V_v$. $f(t)$ is the prediction function of a driving process. The purpose and function of the driving process has been explained in Section 7. Because Active Virtual Network Management Prediction always uses the correct value when the predicted time (τ) equals the current real time (t) and it is assumed that the predictions become more accurate as the predicted time of the event approaches the current time, the reasonable assumption is made that $\lim_{\tau \to t} f(\tau) = V_v$. In order for the Active Virtual Network Management Prediction system to always look ahead, $\forall t \ GVT(t) \geq t$. This means that $\forall n \in \{LPs\}$ and $\forall t \ LVT_{lpn}(t) \geq t$ and $\min_{m \in \{M\}} \{ m \} \geq t$ where m is the receive time of a message, M is the set of messages in the entire system and LVT_{lpn} is the of the nth Logical Process. In other words, the Local Virtual Time of each must be greater than or equal to real time and the smallest message not yet processed must also be greater than or equal to real time. The smallest message could cause a rollback to that time. This implies that $\forall n,t \ LVT_{dpn}(t) \geq t$. In other words, this implies that the Logical Virtual Time of each driving process must be greater than or equal to real time. An out-of-order rollback occurs when $m < LVT(t)$. The largest saved state time such that $t_{SQ} < m$ is used to restore the state of the Logical Process, where t_{SQ} is the real time the state was saved. Then the expected task execution time (τ_{task}) can take no longer than $t_{SQ} - t$ to complete in order for GVT to remain ahead of real time. Thus, a constraint between expected task execution time (τ_{task}), the time associated with a state value (t_{SQ}), and real time (t) has been defined. What remains to be considered is the effect of out-of-tolerance state values on the rollback probability and the concept of stability in Active Virtual Network Management Prediction.

8.2.1 Rollback Rate

Stability in Active Virtual Network Management Prediction is related to the ability of the system to reduce the number of rollbacks. An unstable system is one in which there exist enough rollbacks to cause the system to take longer than real-time to reach the end of the Sliding Lookahead Window. This window has a length of *Lookahead* time units. One end of the window follows the current wallclock time and the other is the distance to which the system should predict. Rollback is caused by the arrival of a message that should have been executed in the past and by out-of-tolerance states. In either case, messages that had been generated prior to the rollback are false messages. Rollback is contained by sending anti-messages to cancel the effects of false messages. The more quickly the anti-messages overtake the effect of false messages, the more efficiently rollback is contained.

One cause of rollbacks in Active Virtual Network Management Prediction is real messages that are out of tolerance. Those processes that require a higher degree of tolerance are most likely to rollback. A worst case probability of out-of-tolerance rollback for a single process, shown in Equation 8.12, is based on Chebycheff's Inequality (Papoulis, 1991) from basic probability. The variance of the data is σ^2 and Θ is the acceptable tolerance for a configuration process. Therefore, the performance gains of Active Virtual Network Management Prediction are reduced as a function of P_{ot}. At the cost of increasing the accuracy of the driving process(es), that is, decreasing σ^2 in Proposition 1, P_{ot} becomes small thus increasing the performance gain of Active Virtual Network Management Prediction.

Proposition 1

The probability of rollback of an LP is

$$P_{ot} \leq \frac{\sigma^2}{\Theta^2} \tag{8.12}$$

where P_{ot} is the probability of out-of-tolerance rollback for an LP, σ^2 is the variance in the amount of error, and Θ is the tolerance allowed for error.

The expected time between rollbacks for the Active Virtual Network Management Prediction system is critical for determining its feasibility. The probability of rollback for all processes is the probability of out-of-order message occurrence and the probability of out-of-tolerance state values ($P_{rb} = P_{oo} + P_{ot}$). The received message rate per is R_m and there are N Logical Processes. The expected inter-rollback time for the system is shown in Equation 8.13.

Proposition 2

The expected inter-rollback time is

$$T_{rb} = \frac{1}{\lambda_{rb}} = \frac{1}{R_m N P_{rb}} \tag{8.13}$$

where T_{rb} is the expected inter-rollback time, λ_{rb} is the expected rollback rate, R_m is the received message rate per , there are N es, and P_{rb} is the probability of rollback per process.

8.2.2 Single Processor Logical Processes

Multiple Logical Processes on a single processor lose any gain in concurrency since they are being served by a single processor; however, the Logical Processes can maintain the Active Virtual Network Management Prediction lookahead if partitioned properly. The single processor logical processes receive virtual messages expected to occur in the future as well as real messages. Because single processor logical processes reside on a single processor, they are not operating in parallel as logical processes do in an optimistic simulation system; thus a new term needs to be applied to a task partitioned into Logical Processes on a single processor. Each partition of tasks into Logical Processes on a single processor is called a Single Processor Logical Process (SLP). In the upper portion of Figure 8.10, a task has been partitioned into two logical processes. The same task exists in the lower portion of Figure 8.10 as a single Logical Process. If task B must rollback because of an out-of-tolerance result, the entire single Logical Process must rollback, while only the Logical Process for task B must rollback in the multiple case. Thus partitioning a task into multiple Logical Processes saves time compared to a single task. Thus, **without considering parallelism**, lookahead is achieved by allowing the sequential system to work ahead while individual tasks within the system are allowed to rollback. Only tasks that deviate beyond a given pre-configured tolerance are rolled back. Thus entire pre-computed and cached results are not lost due to inaccuracy; only parts of pre-computed results must be re-computed. There are significant differences in the behavior of SLP, MLP, and hybrid systems. Each system needs to be analyzed separately.

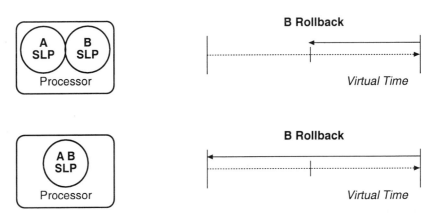

Figure 8.10. Single and Multiple Processor Logical Process System.

Consider the optimal method of partitioning a single processor system into Single Processor Logical Processes in order to obtain speedup over a single process. Assume n tasks, $task_1$, ..., $task_n$, with expected execution times of τ_1, ..., τ_n, and that $task_n$ depends on messages from $task_{n-1}$ with a tolerance of Θ_n. This is the largest error allowed in the input message such that the output is correct. Using the results from Proposition 4.1, it is possible to determine a partitioning of tasks into logical processes such that speedup is achieved over operation of the same tasks encapsulated in a single Logical Process. Figure 8.11 shows possible groupings of the same set of six tasks into logical processes. It is hypothesized that the tasks that are most likely to rollback and those that take the greatest amount of time to execute should be grouped together within Single Processor Logical Processes to minimize the rollback time. There are 2^{n-1} possible groupings of tasks into Single Processor Logical Processes, where n is the number of tasks and message dependency among the tasks is maintained. Those tasks least likely to rollback and those that execute quickly should be grouped within a single Single Processor to reduce the overhead of rollback. For example, if all the tasks in Figure 8.11 have an equal probability of rollback and $\tau_2 >> \max\{\ \tau_1, \tau_3, \ldots\ \}$ then the tasks should be partitioned such that $task_2$ is in a separate Single Processor : ($task_1 \mid task_2 \mid task_3 \ldots task_n$) where "|" indicates the grouping of tasks into sequential logical processes.

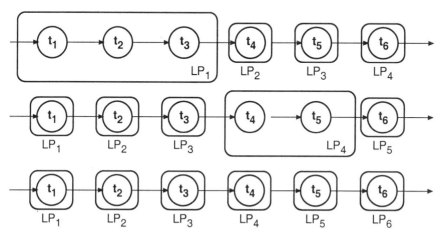

Figure 8.11. Possible Partitioning of Tasks into Logical Processes on a Single Processor.

For example, the expected execution time for five tasks with equal probabilities of rollback of 0.1 are shown in Figure 8.12. It is assumed that these tasks communicate in order starting from Task 1 to Task 5 in order to generate a result. In Figure 8.12, the x-axis indicates the boundary between task partitions as the probability of rollback of task 5 is varied. With an x-value of 3, the solid surface shows the expected execution time for the first three tasks combined within a single and the remainder of the tasks encapsulated in separate Logical Processes. The dashed surface shows the first three tasks encapsulated in separate Logical Processes and the remainder of the tasks encapsulated within a Logical Process. The graph in Figure 8.12 indicates a minimum for both curves when the high probability rollback tasks are encapsulated in separate Logical Processes from the low probability of rollback tasks. As the probability of rollback increases, the expected execution time for all five processes is minimized when Task 5 is encapsulated in a separate Logical Process.

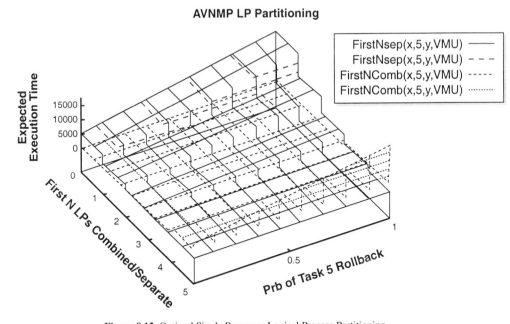

Figure 8.12. Optimal Single Processor Logical Process Partitioning.

8.2.2.1 Task Partition Analysis

Consider an example of Active Virtual Network Management Prediction used for traffic prediction. Assume the computation time is exponentially distributed with mean $[1/(\mu_{r_2})]$. As a simplified example, assume the packet forwarding operation for a router of type A is also exponentially distributed with mean $[1/(\mu_{r_1})]$. The router of type B has a rollback probability of P_{r_2} and takes time τ_{r_2} to rollback. The router of type A has a rollback probability of P_{r_1} and takes time τ_{r_1} to rollback. If both operations are encapsulated by a single logical process, then the expected time of operation is shown in Equation 8.14. If each operation is encapsulated in a separate logical process, then the expected time is shown in Equation 8.15. Equations 8.14 and 8.15 are formed by the sum of the expected time to execute the task, which is the first term, and the rollback time, which is the second term. The probability of rollback in the combined Logical Process is the probability that either task will rollback. Therefore, the expected execution time of the tasks encapsulated in separate Logical Processes is smaller since $\tau_{separate} < \tau_{combined}$.

$$\tau_{combined} = \left(\frac{1}{\mu_{r_2}} + \frac{1}{\mu_{r_1}}\right) + \left(\frac{1}{\mu_{r_2}} + \frac{1}{\mu_{r_1}}\right)\left(P_{r_2} + P_{r_1}\right)\left(\pi_{r_2} + \pi_{r_1}\right) \qquad (8.14)$$

$$\tau_{separate} = \left(\frac{1}{\mu_{r_2}} + \frac{1}{\mu_{r_2}} P_{r_2} \pi_{r_2} \right) + \left(\frac{1}{\mu_{r_1}} + \frac{1}{\mu_{r_1}} P_{r_1} \pi_{r_1} \right) \qquad (8.15)$$

The grouping of tasks into Single Processor Logical Processes can be done dynamically, that is, while the system is in operation. This dynamic adjustment is currently outside the scope of this research but related to optimistic simulation load balancing (Glazer, 1993, Glazer and Tropper, 1993) and the recently developed topic of optimistic simulation dynamic partitioning (Boukerche and Tropper, 1994, Konas and Yew, 1995).

8.2.3 Single Processor Logical Process Prediction Rate

The Local Virtual Time is a particular Logical Process's notion of the current time. In optimistic simulation the Local Virtual Time of individual processes may be different from one another and generally proceed at a much faster rate than real time. Thus, the rate at which a Single Processor system can predict events (prediction rate) is the rate of change of the Single Processor Logical Process's Local Virtual Time with respect to real time. Assume a driving process whose virtual message generation rate is λ_{vm}. The Local Virtual Time is increased by the expected amount Δ_{vm} every $[1/(\lambda_{vm})]$ time units. The expected time spent executing the task is τ_{task}. The random variables X and Y are the proportion of messages that are out-of-order and out-of-tolerance respectively. The expected real time to handle a rollback is τ_{rb}. Then the Single Processor Logical Process's Local Virtual Time advances at the expected rate shown in Proposition 3.

Proposition 3 *[Single Processor Logical Process Speed] The average prediction rate of a single logical processor system is*

$$S_{cache} = \frac{LVT}{t} = \lambda_{vm} \left(\Delta_{vm} - \tau_{task} - (\tau_{task} + \tau_{rb}) E[X] - \left(\Delta_{vm} - \frac{1}{\lambda_{vm}} \right) E[Y] \right) (8.16)$$

where the virtual message generation rate is λ_{vm}, the expected lookahead per message is Δ_{vm}, the proportion of out-of-order messages is X, the proportion of out-of-tolerance messages is Y, τ_{task} is the expected task execution time in real time, τ_{rb} is the expected rollback overhead time in real time, LVT is the Local Virtual Time, and t is real time.

In Proposition 3, the expected lookahead per message (Δ_{vm}) is reduced by the real time taken to process the message (τ_{task}). The expected lookahead is also reduced by the time to re-execute the task (τ_{task}) and the rollback time (τ_{rb}) times the proportion of occurrences of an out-of-order message ($E[X]$) that results in the term ($\tau_{task} + \tau_{rb}$) $E[X]$. Finally, the derivation of the ($\Delta_{vm} - [1/(\lambda_{vm})]$) $E[Y]$ term is shown in Figure 8.13. In Figure 8.13, a real message arrives at time t. Note that real time t and Local Virtual Time are both shown on the same time axis in Figure 8.13. The current Local Virtual Time of the process is labeled at time $LVT(t)$ in Figure 8.13. The dotted line in Figure 8.13 represents the time $\Delta_{vm} - [1/(\lambda_{vm})]$ that is subtracted from the when an out-of-tolerance rollback occurs. The result of the subtraction of $\Delta_{vm} - [1/(\lambda_{vm})]$ from the LVT(t) results in the

Local Virtual Time returning to real time as required by the algorithm. The virtual message inter-arrival time is $[1/(\lambda_{vm})]$. Note that the $(\Delta_{vm} - [1/(\lambda_{vm})])$ $E[Y]$ term causes the speedup to approach 1 based on the frequency of out-of-tolerance rollback ($E[Y]$).

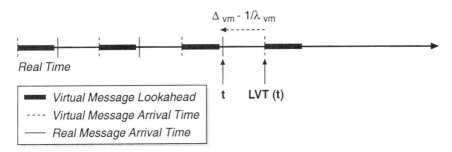

Figure 8.13. Out-of-Tolerance Rollback.

$$\nabla f(x*) + \lambda^T \nabla h(x*) + \mu^T \nabla g(x*) = 0$$
$$\mu^T g(x*) = 0 \qquad\qquad (8.17)$$
$$\mu \geq 0$$

8.2.4 Sensitivity

If the proportion of out-of-tolerance messages, Y, cannot be reduced to zero, the virtual message generation rates and expected virtual message lookahead times can be adjusted in order to improve speedup. Given the closed form expression for Active Virtual Network Management Prediction speedup in Proposition 3, it is important to determine the optimal values for each parameter, particularly λ_{vm} and Δ_{vm} and in addition, the sensitivity of each parameter. Sensitivity information indicates parameters that most affect the speedup. The parameters that most affect the speedup are the ones that yield the best results if optimized.

One technique that optimizes a constrained objective function and that also determines the sensitivity of each parameter within the constraints is the Kuhn-Tucker method (Luenberger, 1989, p. 314). The reason for using this method rather than simply taking the derivative of Equation 8.16 is that the optimal value must reside within a set of constraints. Depending on the particular application of Active Virtual Network Management Prediction, the constraints may become more complex than those shown in this example. The constraints for this example are discussed in detail later. The sensitivity results appear as a by-product of the Kuhn-Tucker method. The first order necessary conditions for an extremum using the Kuhn-Tucker method are listed in Equation 8.17. The second order necessary conditions for an extremum are given in Equation 8.18, where L must be positive semi-definite over the active constraints and L, F, H, and G are Hessians. The second order sufficient conditions are the same as the first order necessary conditions and the Hessian matrix in Equation 8.18 is positive definite on the subspace M = $\{y:\nabla h(x)\ y = 0, \nabla g_j(x)\ y = 0$ for all $j \in J\}$, where $J = \{j: g_j(x) = 0, \mu_j \geq 0\}$. The sensitivity is determined by the Lagrange multipliers, λ^T and μ^T. The Hessian of the objective function and of each of the inequality constraints is a zero matrix; thus, the eigenvalues L in Equation 8.18 are zero and the matrix is clearly positive definite, satisfying both the necessary and sufficient conditions for an extremum.

$$L(x*) = F(x*) + \lambda^T H(x*) + \mu^T G(x*) \tag{8.18}$$

The function f in Equation 8.17 is the Active Virtual Network Management Prediction speedup given in Equation 8.16. The matrix h does not exist, because there are no equality constraints, and the matrix g consists of the inequality constraints that are specified in Equation 8.20.

Clearly the upper bound constraints on $E[X]$ and $E[Y]$ are the virtual message rate. The constraints for τ_{task} and τ_{rb} are based on measurements of the task execution time and the time to execute a rollback. The maximum value for λ_{vm} is determined by the rate at which the virtual message can be processed. Finally, the maximum value for Δ_{vm} is determined by the required caching period. If Δ_{vm} is too large, there may be no state in the SQ with which to compare an incoming real message.

From inspection of Equation 8.16 and the constraint shown in Equation 8.19, the constraints from are $\Delta_{vm} = 45.0$, $\tau_{task} = 5.0$, $\tau_{rb} = 1.0$, $E[X] = 0.0$, $E[Y] = 0.0$ that results in the optimal solution shown in Equation 8.22. The Lagrange multipliers μ_1 through μ_6 show that $E[Y]$ ($-\mu_6 = -8.0$), λ_{vm} ($-\mu_1 = -40.0$), and $E[X]$ ($-\mu_5 = -1.2$) have the greatest sensitivities. Therefore, reducing the out-of-tolerance rollback has the greatest effect on speedup. However, the effect of optimistic synchronization on speedup needs to be studied.

$$l\lambda_{vm} = \cfrac{1}{\tau_{task} + (\tau_{task} + \tau_{rb})E[X] + \left(\Delta_{vm} - \cfrac{1}{\lambda_{vm}}\right)E[Y]} \tag{8.19}$$

$$0.0 \leq \lambda_{vm} \leq \cfrac{1}{\tau_{task} + (\tau_{task} + \tau_{rb})E[X] + \left(\Delta_{vm} - \cfrac{1}{\lambda_{vm}}\right)E[Y]} \tag{8.20}$$

$$0.1 \leq \Delta_{vm} \leq 45.0 \tag{8.21}$$

$$5.0 \leq \tau_{task} \leq 10.0$$
$$1.0 \leq \tau_{rb} \leq 2.0$$
$$0.0 \leq E[X] \leq 1.0$$
$$0.0 \leq E[Y] \leq 1.0$$
$$\lambda_{vm} = 1.0, \mu_1 = 40.0$$
$$\Delta_{vm} = 45.0, \mu_2 = 0.2$$
$$\tau_{task} = 0.0, \mu_3 = 0.2$$
$$\tau_{rb} = 0.0, \mu_4 = 0.0$$
$$E[X] = 0.0, \mu_5 = 1.2$$
$$E[Y] = 0.0, \mu_6 = 8.0$$

$$(8.22)$$

8.2.5 Sequential Execution Multiple Processors

At the time of this writing, a comparison of optimistic synchronization with sequential synchronization cannot be found in the literature because there has been little work on techniques that combine optimistic synchronization and a real time system with the exception of hybrid systems such as the system described in (Bagrodia and Shen, 1991). The hybrid system described in (Bagrodia and Shen, 1991) is used as a design technique in which distributed simulation LPs are gradually replaced with real system components allowing the emulated system to be executed as the system is built. It does not focus on predicting events as in Active Virtual Network Management Prediction. This section examines sequential execution of tasks, which corresponds with non-Active Virtual Network Management Prediction operation as shown in Figure 8.14 in order to compare it with the Active Virtual Network Management Prediction algorithm in the next section. As a specific example, consider K virtual messages with load prediction values passing through P router forwarding processes and each process has an exponential processing time with average $[1/(\mu)]$. In the sequential case, as might be done within the centralized manager as shown in Figure 8.1, the expected completion time should be K times the summation of P exponential distributions. The summation of P exponential distributions is a Gamma Distribution as shown in the sequential execution probability distribution function in Equation 8.23. The average time to complete K tasks is shown in Equation 8.24.

$$f_T(x|P,\mu) = \begin{cases} \dfrac{\mu^P}{\Gamma(P)} x^{P-1} \exp^{-\mu x} & x > 0 \\ 0 & x \leq 0 \end{cases}$$

$$(8.23)$$

$$T_{seq} = K \int_0^\infty x f_T(x|P,\mu)\, dx$$

$$(8.24)$$

Chandy-Misra

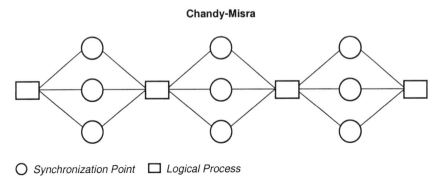

○ *Synchronization Point* □ *Logical Process*

Figure 8.14. Sequential Model of Operation.

8.2.6 Asynchronous Execution Multiple Processors

Assume that an ordering of events is no longer a requirement. This represents the asynchronous Active Virtual Network Management Prediction case and is shown in Figure 8.15. Note that this is the analysis of speedup due to parallelism only, not the lookahead capability of asynchronous Active Virtual Network Management Prediction. This analysis of speedup assumes messages arrive in correct order and thus there is no rollback. However, this also assumes that there are no optimization methods such as lazy cancellation. Following (Felderman and Kleinrock, 1990) the expected completion time is approximated by the maximum of P K-stage Erlangs where P is the number of processes which can execute in parallel at each stage of execution. A K-stage Erlang model represents the total service time as a series of exponential service times, where each service time is performed by a process residing on an independent processor in this case. There is no need to delay processing within the K-stage model because of inter-process dependencies, as there is for synchronous and sequential cases. Equation 8.25 shows the pdf for a K-stage Erlang distribution.

$$f_T(x) = \frac{\mu e^{-\mu x}(\mu x)^{K-1}}{(K-1)!} \tag{8.25}$$

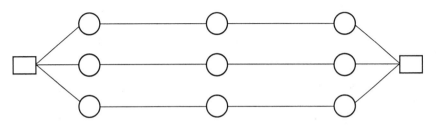

○ *Synchronization Point* □ *Logical Process*

Figure 8.15. Active Virtual Network Management Prediction Model of Parallelism.

As pointed out in (Felderman and Kleinrock, 1990), the probability that a K-stage Erlang takes time less than or equal to t is 1 minus the probability that the K-stage Erlang distribution takes time greater than t, which is simply one minus the probability that there are K arrivals in the interval $[0,t]$ from a Poisson process at rate μ. This result is shown in Equation 8.26.

$$F_T(x) = 1 - e^{-\mu x} \sum_{i=0}^{K-1} \frac{(\mu x)^i}{i!} \qquad (8.26)$$

$$T_{async} = \int_0^\infty [1 - F_T(x)] dx \qquad (8.27)$$

The expected value is shown in Equation 8.27. This integral is hard to solve with a closed form solution and (Felderman and Kleinrock, 1990) instead try to find an approximate equation. This study attempts to be exact by using Equation 8.27 and solving it numerically (Kleinrock, 1975, p. 378). In Equation 8.28 $S_{parallel}$ is the speedup of optimistic synchronization over strictly sequential synchronization and is graphed in Figure 8.16 as a function of the number of processors. The speedup gained by parallelism ($S_{parallel}$) augments the speedup due to lookahead (S_{cache}) as shown in Equation 8.29, where the (PR) is the Active Virtual Network Management Prediction speedup and X and Y are random variables representing the proportion of out-of-order and out-of-tolerance messages respectively.

$$S_{parallel} = \frac{T_{seq}}{T_{async}} \qquad (8.28)$$

$$PR_{X,Y} = \lambda_{vm}\left(\Delta_{vm}S_{parallel} - \tau_{task} - (\tau_{task} + \tau_{rb})X - \left(\Delta_{vm}S_{parallel} - \frac{1}{\lambda_{vm}}\right)Y\right)$$

$$(8.29)$$

There is clearly a potential speedup in Active Virtual Network Management Prediction in contrast to a single processor model of the network. The Active Virtual Network Management Prediction algorithm implementation is able to take advantage of both Single Processor Logical Processes (Slogical Process) lookahead without parallel processing and speedup due to parallel processing because Active Virtual Network Management Prediction has been implemented on many nodes throughout the network and each node has its own processor. Note that while Clustered Time Warp (Avril, 1996), which was developed concurrently but independently of Active Virtual Network Management Prediction, uses a similar concept to Single Processor Logical Processes and Logical Process, it does not consider a real-time system as in Active Virtual Network Management Prediction.

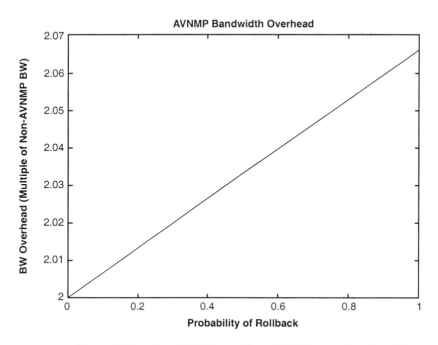

Figure 8.16. Speedup of AVNMP over Non-AVNMP Systems Due to Parallelism.

8.2.7 Multiple Processor Logical Processes

The goal of Active Virtual Network Management Prediction is to provide accurate predictions quickly enough so that the results are available before they are needed. Without taking advantage of parallelism, a less sophisticated algorithm than Active Virtual Network Management Prediction could run ahead of real-time and cache results for future use. This is done in the Sequential Processor system, which assumes strict synchronization between processes whose prediction rate is defined in Proposition 3. With such a simpler mechanism, P_{oo} and $E[X]$ are always zero. However, simply predicting and caching results ahead of time does not fully utilize inherent parallelism in the system as long as messages between Logical Processes remain strictly synchronized. Strict synchronization means that processes must wait until all messages are insured to be processed in order. Any speedup to be gained through parallelism comes from the same mechanism as in optimistic parallel simulation; the assumption that messages arrive in order by TR, thus eliminating unnecessary synchronization delay. However, messages arrive out-of-order in Active Virtual Network Management Prediction for the following reasons. A general-purpose system using the Active Virtual Network Management Prediction algorithm may have multiple driving processes, each predicting at different rates into the future. Another reason for out-of-order messages is that Logical Processes are not required to wait until processing completes before sending the next message.

Also, processes may run faster for virtual computations by allowing a larger tolerance. Finally, for testing purposes, hardware or processes may be replaced with simulated code, thus generating results faster than the actual process would. Thus, although real and future time are working in parallel with strict synchronization, no advantage is being taken of parallel processing. This is demonstrated by the fact that, with strict synchronization of messages, the same speedup (S_{cache}) as defined in Proposition 3 occurs regardless of whether a single processor or multiple processors are used. What differentiates Active Virtual Network Management Prediction is the fact that it takes advantage of inherent parallelism in the system as compared to a sequential non-Active Virtual Network Management Prediction pre-computation and caching method. Thus it is better able to meet the deadline imposed by predicting results before they are required. To see why this is true, consider what happens as the overhead terms in Proposition 3, τ_{task} $- (\tau_{task} + \tau_{rb})\ E[X] - (\Delta_{vm}S_{parallel} - [1/(\lambda_{vm})])\ E[Y]$, approach Δ_{vm}. The prediction rate becomes equal to real-time and can fall behind real-time as $\tau_{task} - (\tau_{task} + \tau_{rb})\ E[X] - (\Delta_{vm}S_{parallel} - [1/(\lambda_{vm})])\ E[Y]$ becomes larger. Optimistic synchronization helps to alleviate the problem of the prediction rate falling behind real-time. Optimistic synchronization has another advantageous property, super-criticality. A super critical system is one that can compute results faster than the time taken by the critical path through the system. This can occur in Active Virtual Network Management Prediction using the lazy cancellation optimization as discussed in Section 7. Super-criticality occurs when task execution with false message values generates a correct result. Thus prematurely executed tasks do not rollback and a correct result is generated faster than the route through the critical path.

The Active Virtual Network Management Prediction algorithm has two forms of speedup that need to be clearly defined. There is the speedup in availability of results because they have been pre-computed and cached. There is also the speedup due to more efficient usage of parallelism. The gain in speedup due to parallelism in Active Virtual Network Management Prediction can be significant given the proper conditions. In order to prevent confusion about the type of speedup being analyzed, the speedup due to pre-computing and caching results is defined as S_{cache} and the speedup due to parallelism is defined as $S_{parallel}$. Speedup due to parallelism among multiple processors in Active Virtual Network Management Prediction is gained from the same mechanism that provides speedup in parallel simulation, that is, it is assumed that all relevant messages are present and are processed in order by receive time. The method of maintaining message order is optimistic in the form of rollback. The following sections look at $S_{parallel}$ due to a multiprocessor configuration system.

8.2.8 AVNMP Prediction Rate with a Fixed Lookahead

There are three possible cases to consider when determining the speedup of Active Virtual Network Management Prediction over non-lookahead sequential execution. The speedup given each of these cases and their respective probabilities needs to be analyzed. These cases are illustrated in Figures 8.17 through 8.19. The time that an event is predicted to occur and the result cached is labeled $t_{virtual\ event}$, the time a real event occurs is labeled $t_{real\ event}$, and the time a result for the real event is calculated is labeled $t_{no-avnmp}$. In Active Virtual Network Management Prediction, the virtual event and its result can be cached before the real event, as shown in Figure 8.17, between the real event and the time the real event result is calculated as shown in Figure 8.18, or after the real event result is calculated as shown in Figure 8.19. In each case, all events are considered relative to the occurrence of the real event. It is assumed that the real event occurs at time t. A random variable called the lookahead (LA) is defined as $LVT - t$. The virtual event occurs at time $t - LA$. Assume that the task that must be executed once the real event occurs takes τ_{task} time. Then without Active Virtual Network Management Prediction the task is completed at time $t + \tau_{task}$.

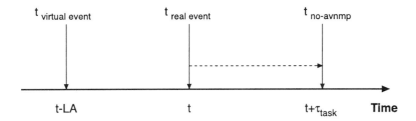

Figure 8.17. AVNMP Prediction Cached before Real Event.

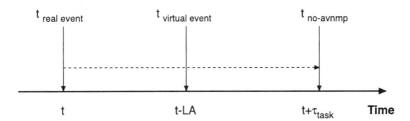

Figure 8.18. AVNMP Prediction Cached Later than Real Event.

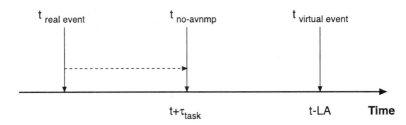

Figure 8.19. AVNMP Prediction Cached Slower than Real Time.

The prediction rate has been defined in Equation 8.29 and includes the time to predict an event and cache the result in the State Queue. Recall that in Section 2 the expected value of X has been determined based on the inherent synchronization of the topology. It was shown that X has an expected value that varies with the rate of hand-offs.

It is clear that the proportion of out-of-order messages is dependent on the architecture and the partitioning of tasks into Logical Processes. Thus, it is difficult in an experimental implementation to vary X. It is easier to change the tolerance rather than change the architecture to evaluate the performance of Active Virtual Network Management Prediction. For these reasons, the analysis proceeds with $PR_{X,Y|X\,=\,E[X]}$. Since the prediction rate is the rate of change of Local Virtual Time with respect to time, the value of the Local Virtual Time is shown in Equation 8.30, where C is an initial offset. This offset may occur because Active Virtual Network Management Prediction may begin running C time units before or after the real system. Replacing LVT in the definition of LA with the right side of Equation 8.30 yields the Equation for lookahead shown in Equation 8.31.

$$LVT_{X,Y|x=E[X]} = \lambda_{vm}\left(\Delta_{vm}S_{parallel} - \tau_{task} - (\tau_{task} + \tau_{rb})E[X] - \left(\Delta_{vm}S_{parallel} - \frac{1}{\lambda_{vm}} \right)Y \right)t + C$$

(8.30)

$$LA_{X,Y|X=E[X]} = \left(LVT_{X,Y|X=E[X]} - 1 \right) + C \qquad (8.31)$$

The probability of the event in which the Active Virtual Network Management Prediction result is cached before the real event is defined in Equation 8.32. The probability of the event for which the Active Virtual Network Management Prediction result is cached after the real event but before the result would have been calculated in the non-Active Virtual Network Management Prediction system is defined in Equation 8.33. Finally, the probability of the event for which the Active Virtual Network Management Prediction result is cached after the result would have been calculated in a non-Active Virtual Network Management Prediction system is defined in Equation 8.34.

$$P_{cache} = P\left[LA_{X,Y|X=E[X]} > \tau_{task} \right] \qquad (8.32)$$

$$P_{late} = P\left[0 \le LA_{X,Y|X=E[X]} \le \tau_{task} \right] \qquad (8.33)$$

$$P_{slow} = P\left[LA_{X,Y|X=E[X]} < 0 \right] \qquad (8.34)$$

The goal of this analysis is to determine the effect of the proportion of out-of-tolerance messages (Y) on the speedup of an Active Virtual Network Management Prediction system. Hence we assume that the proportion Y is a binomially distributed random variable with parameters n and p where n is the total number of messages and p is the probability of any single message being out of tolerance. It is helpful to simplify Equation 8.31 by using γ_1 and γ_2 as defined in Equations 8.36 and 8.37 in Equation 8.35.

$$LA_{X,Y|X=E[X]} = \gamma_1 - \gamma_2 Y \qquad (8.35)$$

$$\gamma_1 = \left(\lambda_{vm}\Delta_{vm}S_{parallel} - \lambda_{vm}\tau_{task} - \lambda vm(\tau_{rb} + \tau_{task})E[X] - 1\right)t + C \qquad (8.36)$$

$$\gamma_2 = \lambda_{vm}\left(\Delta_{vm}S_{parallel} - \frac{1}{\lambda_{vm}} + \tau_{rb}\right)t \qquad (8.37)$$

The early prediction probability as illustrated in Figure 8.17 is shown in Equation 8.38. The late prediction probability as illustrated in Figure 8.18 is shown in Equation 8.38. The probability for which Active Virtual Network Management Prediction falls behind real time as illustrated in Figure 8.19 is shown in Equation 8.40. The three cases for determining Active Virtual Network Management Prediction speedup are thus determined by the probability that Y is greater or less than two thresholds.

$$P_1(t) = P_{cache} \quad X,Y|X=E[X] = P\left[Y < \frac{\gamma_1 - \tau_{task}}{\gamma_2}\right] \qquad (8.38)$$

$$P_2(t) = P_{late} \quad X,Y|X=E[X] = P\left[\frac{\gamma_1 - \tau_{task}}{\gamma_2} \leq Y \frac{\gamma_1}{\gamma_2}\right] \qquad (8.39)$$

$$P_3(t) = P_{slow} \quad X,Y|X=E[X] = P\left[Y > \frac{\gamma_1}{\gamma_2}\right] \qquad (8.40)$$

The three probabilities in Equations 8.38 through 8.40 depend on (Y) and real time because the analysis assumes that the lookahead increases indefinitely, which shifts the thresholds in such a manner as to increase Active Virtual Network Management Prediction performance as real time increases. However, the Active Virtual Network Management Prediction algorithm holds processing of virtual messages once the end of the Sliding Lookahead Window is reached. The hold time occurs when $LA = \Lambda$ where Λ is the length of the Sliding Lookahead Window. Once Λ is reached, processing of virtual messages is discontinued until real-time reaches Local Virtual Time. The lookahead versus real time including the effect of the Sliding Lookahead Window is shown in Figure 8.20. The dashed arrow represents the lookahead which increases at rate PR. The solid line returning to zero is lookahead as the Logical Process delays. Because the curve in Figure 8.20 from 0 to t_L repeats indefinitely, only the area from 0 to t_L need be considered. For each $P_i(t)$ $i = 1,2,3$, the time average over the lookahead time (t_L) is shown by the integral in Equation 8.41.

$$P_{X,Y|X=E[X]} = \frac{1}{t_L}\int_0^{t_L} P_i t \,(dt) \qquad (8.41)$$

$$\eta \equiv P_{cache\ X|X=E[X]}C_r + \left(P_{late\ X|X=E[X]} + P_{slow\ X|X=E[X]}\right)PR_{X,Y|X=E[X]}$$

(8.42)

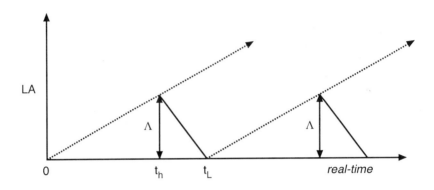

Figure 8.20. Lookahead with a Sliding Lookahead Window.

The probability of each of the events shown in Figures 8.17 through 8.19 is multiplied by the speedup for each event in order to derive the average speedup. For the case shown in Figure 8.17, the speedup (C_r) is provided by the time to read the cache over directly computing the result. For the remaining cases the speedup is $PR_{X,Y|X = E[X]}$ that has been defined as $[(LVT_{X,Y|X = E[X]})/t]$ as shown in Equation 8.42. The analytical results for speedup are graphed in Figure 8.21. A high probability of out-of-tolerance rollback in Figure 8.21 results in a speedup of less than one. Real messages are always processed when they arrive at a Logical Process. Thus, no matter how late Active Virtual Network Management Prediction results are, the system continues to run near real time. However, when Active Virtual Network Management Prediction results are very late due to a high proportion of out-of-tolerance messages, the Active Virtual Network Management Prediction system is slower than real time because out-of-tolerance rollback overhead processing occurs. Anti-messages must be sent to correct other Logical Processes that have processed messages which have now been found to be out of tolerance from the current Logical Process. This causes the speedup to be less than one when the out-of-tolerance probability is high. Thus, $PR_{X,Y|X = E[X]}$ will be less than one for the "slow" predictions shown in Figure 8.19.

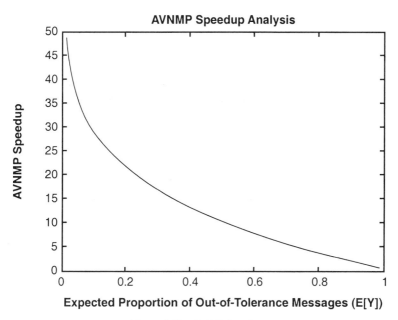

Figure 8.21. AVNMP Speedup.

8.3 PREDICTION ACCURACY

This section derives the prediction accuracy and bandwidth overhead of AVNMP and uses these relationships along with the expected speedup from the previous section to analyze the performance of AVNMP.

8.3.1 Prediction of Accuracy: α

Accuracy is the ability of the system to predict future events. A higher degree of accuracy will result in more "cache hits" of the predicted state cache information. Smaller tolerances should result in greater system accuracy, but this comes at the cost of a reduction in speedup.

Assume for simplicity that the effects of non-causality are negligible for the analysis in this section. The effects of causality are discussed in more detail in Section 8.2. A Logical Process may deviate from the real object it represents either because the Logical Process does not accurately represent the actual entity or because events outside the scope of the predictive system may effect the entities being managed. Ignore events outside the scope of the predictive system for this analysis and consider only the deterministic error from inaccurate prediction of the driving process. The error is defined as the difference between an actual message value at the current time (v_t) and a message value that had been predicted earlier (v_{tp}). Thus the **Message Error** is $ME = v_t - v_{tp}$. Virtual message values generated from a driving process may contain some error. It is assumed that the error in any output message generated by a process is a function of any error in the input message and the amount of time it takes to process the message. A larger processing time increases the chances that external events may have changed before the processing has completed.

Two functions of total **AC**cumulated message value error (AC($\cdot$)) in a predicted result are described by Equations 8.43 and 8.44 and are illustrated in Figure 8.22. ME_{lp0} is the amount of error in the value of the virtual message injected into the predictive system by the driving process (lp_o). The error introduced into the value of the output message produced by the computation of each is represented by the Computation Error function $CE_{lpn}(ME_{lpn-1}, t_{lpn})$. The real time taken for the nth Logical Process to generate a message is t_{lpn}. The error accumulates in the State Queue at each node by the amount $CE_{lpn}(ME_{lpn-1}, t_{lpn})$, which is a function of the error contained in the input message from the predecessor and the time to process that message. Figure 8.22 shows a driving process (DP) generating a virtual message that contains prediction error (ME_{lp0}). The virtual message with prediction error (ME_{lp0}) is processed by node LP_1 in t_{lp1} time units resulting in an output message with error, $ME_{lp1} = CE_{lp0}(ME_{lp0}, t_{lp1})$.

Proposition 4

The accumulated error in a message value is Equation 8.43 and Equation 8.44.

$$AC_n(n) = \sum_{i=1}^{N} CE_{lp_i}\left(ME_{lp_{i-1}}, t_{lp_i}\right) \tag{8.43}$$

$$AC_t(\tau) = \lim_{\sum t_{lp_i} \to \tau} \sum_{i=1}^{n} CE_{lp_i}\left(ME_{lp_{i-1}}, t_{lp_i}\right) \tag{8.44}$$

$$\alpha \equiv \Pr\left[\tan^{-1}\left(\frac{\sigma}{d}\right) > \Theta\right] \tag{8.45}$$

Where Ce_{lpi} is the computational error added to a virtual message value, ME_{lpi} is the virtual message input error, and $t_{lp}i$ is the real time taken to process a virtual message.

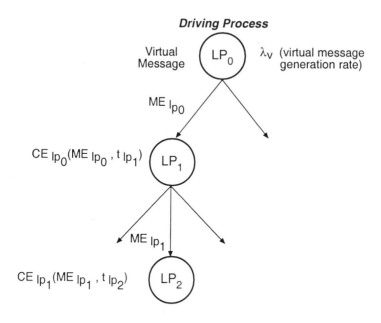

Figure 8.22. Accumulated Message Value Error.

As shown in Proposition 4, $AC_n(n)$ is the total accumulated error in the virtual message output by the nth from the driving process. $AC_t(\tau)$ is the accumulated error in τ real time units from the generation of the initial virtual message from the driving process. Equation is $\lim_{\Sigma_{tlpi}} \to \tau \sum_{i=1}^{n} AC_n(n)$, where n is the number of computations in time τ. In other words, $AC_t(\tau)$ is the error accumulated as messages pass through n Logical Processes in real time τ. For example, if a prediction result is generated in the third Logical Process from the driving process, then the total accumulated error in the result is $AC_n(3)$. If 10 represents the number of time units after the initial message was generated from the driving process, then $AC_t(10)$ would be the amount of total accumulated error in the result. A cache hit occurs when $|AC_t(\tau)| \leq \Theta$, where Θ is the tolerance associated with the last Logical Process required to generate the final result. Equations (8.43) and (8.44) provide a means of representing the amount of error in an Active Virtual Network Management Prediction generated result. Once an event has been predicted and results pre-computed and cached, it would be useful to know what the probability is that the result has been accurately calculated, especially if any results are committed before a real message arrives. The out-of-tolerance check and rollback does not occur until a real message arrives. If a resource is allocated ahead of time based on the predicted result, then this section has defined $\alpha = P[|AC_t(\Lambda)| > \Theta]$ where Θ is the Active Virtual Network Management Prediction tolerance associated with the last Logical Process required to generate the final result.

8.3.2 Bandwidth: β

The amount of overhead in bandwidth required by Active Virtual Network Management Prediction is due to virtual and anti-message load. With perfect prediction capability, there should be exactly one virtual message from the driving process for each real message. The inter-rollback time, $[1/(\lambda_{rb})]$, has been determined in Proposition 3, Equation 8.13. Virtual messages are arriving and generating new messages at a rate of λ_v. Thus, the worst case expected number of messages in the State Queue that will be sent as anti-messages is $[(\lambda_v)/(\lambda_{rb})]$ when a rollback occurs. The bandwidth overhead is shown in Equation 8.46, where λ_v is the virtual message load, λ_r is the real message load, and λ_{rb} is the expected rollback rate. The bandwidth overhead as a function of rollback rate is shown in Figure 8.23. Scalability in Active Virtual Network Management Prediction is the rate at which the proportion of rollbacks increases as the number of nodes increases. The graph in Figure 8.24 illustrates the tradeoff between the number of Logical Processes and the rollback rate given $\lambda_{vm} = 0.03$ virtual messages per millisecond, $\Delta_{vm} = 30.0$ milliseconds, $\tau_{task} = 7.0$ milliseconds, $\tau_{rb} = 1.0$ milliseconds, $S_{parallel} = 1.5$ and $C_r = 100$ where C_r is the speedup gained from reading the cache over computing the result and Rm $= [2/30$ ms$]$. The rollback rate in this graph is the sum of both the out-of-order and the out-of-tolerance rollback rates.

Proposition 5

The expected bandwidth overhead is

$$\beta = \frac{\dfrac{\lambda_v}{\lambda_{rb}} + \lambda_v + \lambda_r}{\lambda_r} \qquad (8.46)$$

where λ_{rb} is the expected rollback rate, λ_v is the expected virtual message rate, and λ_r is the expected real message rate.

8.3.3 Analysis of AVNMP Performance

Equation 8.47 shows the complete Active Virtual Network Management Prediction performance utility. The surface plot showing the utility of Active Virtual Network Management Prediction as a function of the proportion of out-of-tolerance messages is shown in Figure 8.25 where Φ_s, Φ_w, Φ_b are one and $\lambda_{vm} = 0.03$ virtual messages per millisecond, $\Delta_{vm} = 30.0$ milliseconds, $\tau_{task} = 7.0$ milliseconds, $\tau_{rb} = 1.0$ milliseconds, $S_{parallel} = 1.5$ and $C_r = 100$ where C_r is the speedup gained from reading the cache over computing the result. The wasted resources utility is not included in Figure 8.25 because there is only one level of message generation and thus no error accumulation. The y-axis is the relative marginal utility of speedup over reduction in bandwidth overhead SB = $[(\Phi_s)/(\Phi_b)]$. Thus if bandwidth reduction is much more important than speedup, the utility is low and the proportion of rollback messages would have to be kept below 0.3 per millisecond in this case. However, if speedup is the primary desire relative to bandwidth, the proportion of out-of-tolerance rollback message values can be as high as 0.5 per millisecond. If the proportion of out-of-tolerance messages becomes too high, the utility becomes negative because prediction time begins to fall behind real time.

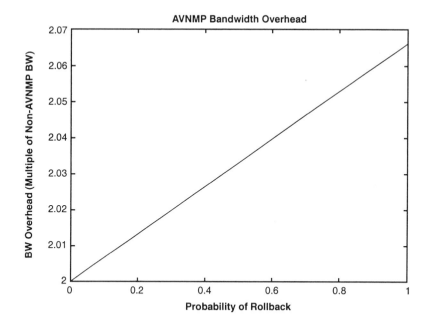

Figure 8.23. AVNMP Bandwidth Overhead.

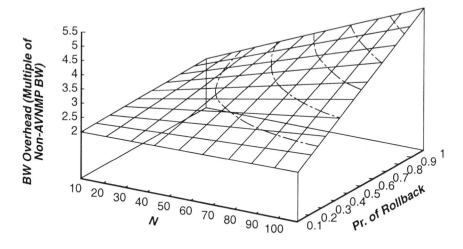

Figure 8.24. AVNMP Scalability.

The effect of the proportion of out-of-order and out-of-tolerance messages on Active Virtual Network Management Prediction speedup is shown in Figure 8.26. This graph shows that out-of-tolerance rollbacks have a greater impact on speedup than out-of-order rollbacks. The reason for the greater impact of the proportion of out-of-tolerance messages is that such rollbacks caused by such messages always cause a process to rollback to real time. An out-of-order rollback only requires the process to rollback to the previous saved state.

Figure 8.27 shows the effect of the proportion of virtual messages and expected lookahead per virtual message on speedup. This graph is interesting because it shows how the proportion of virtual messages injected into the Active Virtual Network Management Prediction system and the expected lookahead time of each message can affect the speedup. The real and virtual message rates are [0.1/ms], Rm = [2/30 ms], λ_{vm} = 0.03 virtual messages per millisecond, Δ_{vm} = 30.0 milliseconds, τ_{task} = 7.0 milliseconds, τ_{rb} = 1.0 milliseconds, $S_{parallel}$ = 1.5 and C_r = 100 where C_r is the speedup gained from reading the cache over computing the result.

$$U_{AVNMP} = \left(P_{cache\ X|X=E[X]}C_r + \left(P_{late\ X|X=E[X]} + P_{slow\ X|X=E[X]}\right)PR_{X,Y|X=E[X]}\right)$$

$$\Phi_s - P[|AC_t(\Lambda)| > \Theta]\Phi_w - \left(\frac{\frac{\lambda_v}{\lambda_{rb}} + \lambda_v}{\lambda_r}\right)\Phi_b$$

$$(8.47)$$

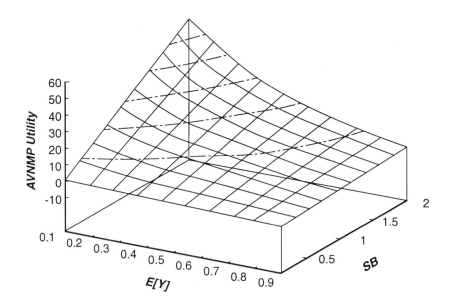

Figure 8.25. Overhead versus Speedup as a Function of Probability of Rollback.

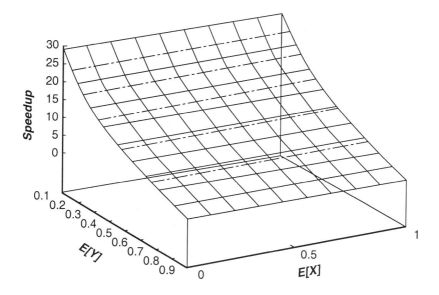

Figure 8.26. Effect of Non-Causality and Tolerance on Speedup.

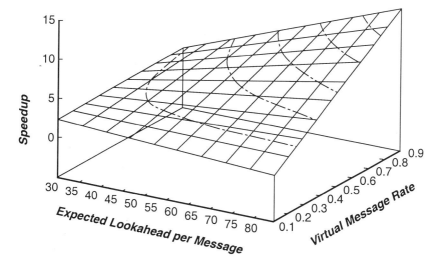

Figure 8.27. Effect of Virtual Message Rate and Lookahead on Speedup.

8.4 EXERCISES

1. Consider a Driving Process whose accuracy improves based on past experience. Suppose that the rate of improvement is $\iota(t)$ where ι is the operational lifetime of the Driving Process. Derive the AVNMP performance and overhead equations using this new information.

2. Suppose multiple future events are incorporated into the AVNMP system. Derive an equation to quantify the probability of occurrence of each event.

3. Consider hardware/software faults of system components. Identify and quantify the effect of component failures on the AVNMP system. How can AVNMP inaccuracy be distinguished from system failure? Will AVNMP adapt to system failures?

4. In regards to the Active Network Management Framework and concepts involving self-management, can AVNMP optimize its own parameter settings, e.g., λ_{vm}, Δ_{vm}, LA?

9

ASPECTS OF AVNMP PERFORMANCE

The following sections discuss other aspects and optimizations of the Active Virtual Network Management Prediction algorithm including handling multiple future events and the relevance of Global Virtual Time to Active Virtual Network Management Prediction. Since all possible alternative events cannot be predicted, only the most likely events are predicted in Active Virtual Network Management Prediction. However, knowledge of alternative events with a lower probability of occurrence allow the system to prepare more intelligently.

Another consideration is the calculation of Global Virtual Time. This requires bandwidth and processing overhead. A bandwidth optimization is suggested in which real packets may be sent less frequently.

9.1 MULTIPLE FUTURE EVENTS

The architecture for implementing alternative futures discussed in Section 7, while a simple and natural extension of the Active Virtual Network Management Prediction algorithm creates additional messages and increases message sizes. Messages require an additional field to identify the probability of occurrence and an event identifier. However, the Active Virtual Network Management Prediction tolerance is shown to provide consideration of events that fall within the tolerances Θ_n where n $\in$ N and N is the number of Logical Processes.

The set of possible futures at time t is represented by the set E. A message value generating an event occurring in one of the possible futures is represented by E_{val}. As messages propagate through the Active Virtual Network Management Prediction system, there is a neighborhood around each message value defined by the tolerance (Θ_n). However, each message value also accumulates error $(AC_n(n))$. Let the neighborhood (E_Δ) be defined such that $E_\Delta \leq |\Theta_n - AC_n(n)|$ for each $n \in \{LPs\}$. Thus, $|E_\Delta + AC_n(n)| \leq \min_{n \in N} \Theta_n$ defines a valid prediction. The infinite set of events in the neighborhood $E_\Delta \leq |\min_{n \in N} \Theta_n - AC_n(n)|$ are valid. Therefore, multiple future events that fall within the bounds of the tolerances reduced by any accumulated error can be implicitly considered.

9.2 GLOBAL VIRTUAL TIME

In order to maintain the lookahead (Λ), for the entire configuration system, it is necessary to know how far into the future the system is currently predicting. The purpose of Global Virtual Time is to determine Λ where Λ is used to stop the Active Virtual Network Management Prediction system from looking ahead once the system has predicted up to the lookahead time. This helps maintain synchronization and saves processing and bandwidth since it is not necessary to continue the prediction process indefinitely into the future, especially since the prediction process is assumed to be less accurate the further it predicts into the future.

Distributed simulation mechanisms require Global Virtual Time in order to determine when to commit events. This is because the simulation cannot rollback beyond Global Virtual Time. In Active Virtual Network Management Prediction, event results are assumed to be cached before real time reaches the Local Virtual Time of a Logical Process. The only purpose for Global Virtual Time in Active Virtual Network Management Prediction is to act as a throttle on computation into the future. Thus, the complexity and overhead required to accurately determine the Global Virtual Time is unnecessary in Active Virtual Network Management Prediction. In the Active Virtual Network Management Prediction system, while the Local Virtual Time of a Logical Process is greater than $t + \Lambda$, the Logical Process does not process virtual messages.

The Global Virtual Time update request packets have the intelligence to travel only to those logical processes most likely to contain a global minimum. An example is shown in Figure 9.1.

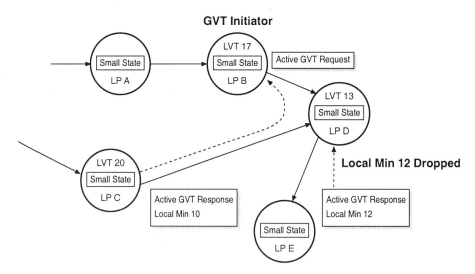

Figure 9.1. Active Global Virtual Time Calculation Overview.

The Active Request packet notices that the logical process with a Global Virtual Time of 20 is greater than the last logical process that the Active Request packet passed through and thus destroys itself. This limits the amount of unnecessary traffic and computation. The nodes that receive the Active Request packet forward the result to the initiator. As the Active Response packets return to the initiator, the last packet is maintained in the cache of each logical process. If the value of the Active Response packet is greater than or equal to the value in the cache, then the packet is dropped. Again, this reduces the amount of traffic and computation that must be performed.

9.3 REAL MESSAGE OPTIMIZATION

Real messages are only used in the Active Virtual Network Management Prediction algorithm as a verification that a prediction has been accurate within a given tolerance. The driving process need not send a real message if the virtual messages are within the lowest tolerance in the path of a virtual message. This requires that the driving process have knowledge of the tolerance of the destination process. The driving process has copies of previously sent messages in its send queue. If real messages are only sent when an out-of-tolerance condition occurs, then the bandwidth can be reduced by up to 50%. Figure 9.2 compares the bandwidth with and without the real message optimization.

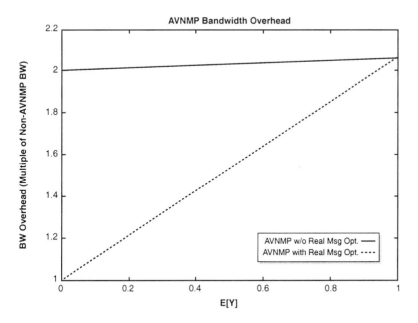

Figure 9.2. Bandwidth Overhead Reduction.

The performance analysis of Active Virtual Network Management Prediction has quantified the costs versus the speedup provided by Active Virtual Network Management Prediction. The costs have been identified as the additional bandwidth and possible wasted resources due to inaccurate prediction. Since the Active Virtual Network Management Prediction algorithm combines optimistic synchronization with a real time system, the probability of non-causal message order was determined. A new approach using Petri-Nets and synchronic distance determined the likelihood of out-of-order virtual messages. The speedup was defined as the expected rate of change of the Local Virtual Time with respect to real time. The speedup was quantified and a sensitivity analysis revealed the parameters most affecting speedup. The bandwidth was quantified based on the probability of rollback and the expected rollback rate of the Active Virtual Network Management Prediction system. A general analysis of the accumulated error of the Active Virtual Network Management Prediction system followed with the probability of error in the active network. Finally, the consideration of alternative future events, the relevance of Global Virtual Time, and a bandwidth technique were discussed.

Active networks enable an exciting new paradigm for communications. This paradigm facilitates the use of data transmission and computation in ways unimaginable in legacy networks. Hopefully the information provided in this book will give the reader a running start in understanding this new technology and generate new ideas in the reader's mind for novel applications of this technology.

9.4 COMPLEXITY IN SELF-PREDICTIVE SYSTEMS

A fascinating perspective on the topic of self-predictive systems is found in *Gödel, Escher, and Bach: An Eternal Golden Braid*, which is a wonderful look at the nature of Human and Artificial Intelligence. A central point in (Hofstadter, 1980) is that intelligence is a Tangled Hierarchy, illustrated in the famous Escher drawing of two hands – each drawing the other. A hand performing the act of drawing is expected to be a level above the hand being drawn. When the two levels are folded together, a Tangled Hierarchy results, an idea which is expressed much more elegantly in (Hofstadter, 1980). Active Virtual Network Management Prediction as presented in this work is a Tangled Hierarchy on several levels: simulation-reality and also present-future time. One of the hands in the Escher drawing represents prediction based on simulation and the other represents reality, each modifying the other in the Active Virtual Network Management Prediction algorithm. However, there is a much deeper mathematical relationship present in this algorithm that relates to Gödel's Theorem. In a nutshell, Gödel's theorem states that no formal system can describe itself with complete fidelity. This places a formidable limitation on the ability of mathematics to describe itself. The implication for artificial intelligence is that the human mind can never fully understand its own operation, or possibly that if one could fully understand how one thinks while one is thinking, then one would cease to "be." In the much more mundane Active Virtual Network Management Prediction algorithm, a system is in some sense attempting to use itself to predict its own future state with the goal of perfect fidelity. If Gödel's Theorem applies, then perfect fidelity is an impossible goal. However, by allowing for a given tolerance in the amount of error and assuming accuracy in prediction which increases as real time approaches the actual time of an event, this study assumes that a useful self-predictive system can be implemented.

In the course of efforts to fully utilize the power of active networks to build a self-managing communications network, the nature of entanglement and the relationship between modeling and communication becomes of utmost importance. This section provides a general overview of the goal that Active Virtual Network Management Prediction is trying to accomplish as well as its evolution as resources increase; that is, how does such a self-predictive system behave as processing and bandwidth become ever larger and more powerful. An attempt is made to identify new theories required to understand such highly self-predictive systems.

9.4.1 Near-Infinite Resources

Now, imagine stepping across a discontinuity into a world where computing power, bandwidth, and computational ubiquity are nearly infinite. Our vision focuses on effects that near-perfect self-prediction would have upon such a world. First we would have near-perfect optimization of resources since local minima could be pushed far into the horizon. Second, currently wasted effort could be avoided, since the outcome of any action could be determined with very precise limits. Critical missing elements are a theory and applications involving highly predictive systems and components. Further study is needed to explore the exciting new world of near-perfect self-prediction and the relationship between highly predictive systems and communications in particular. Figure 9.3 shows an abstract view of computers embedded within almost all devices. Current engineering organizes computing devices in such a way as to optimize communications performance. In our hypothetical world of near-perfect predictive capabilities, direct communication is less important and, in many cases, no longer required, as discussed later. Instead, computational organization is based on forming systems or islands of near-perfect self-prediction. As shown in Figure 9.4, self-predictive capability is used to enhance the performance of the system, which in turn improves the predictive capability, which again improves the performance of the system, ad infinitum, driving the error towards zero.

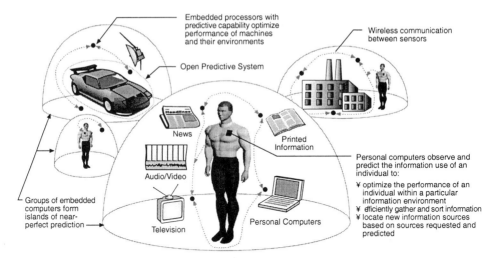

Figure 9.3. Computational organization is based on forming systems or islands of near-perfect self-prediction.

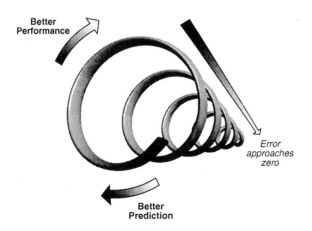

Figure 9.4. This predictive capability is used to drive the error toward zero.

Why do we assume rather than perfect prediction and why do we assume islands rather than perfect prediction everywhere? Clearly, perfect prediction everywhere would take us into a deterministic world where the final outcome of all choices would be known to everyone and the optimal choice could be determined in all cases. In this project it is assumed that limits, however small, exist, such as lack of knowledge about quantum state or of the depths of space. In order to study near-perfect self-predictive islands, the characteristics of such islands need to be identified. It would appear that closed self-predictive islands would be the easiest to understand. The scope of closed self-predictive islands includes all driving forces acting upon the system. Imagine that one has full knowledge of the state of a room full of ping-pong balls and their elasticity. This information can be used to predict the position of the balls at any point in time. However, one is external to the room. The goal is for the balls to predict their own behavior as illustrated in the inner sphere of Figure 9.5. If elasticity represents the dynamics of communication endpoint entities A and B, and movement of the ping-pong balls represents communication, then any exchange of information between A and B is unnecessary since it can be perfectly predicted. Instead of transmitting messages between A and B, an initial transmission of the dynamics of A and B is transmitted to each other, perhaps as active packets within an active network environment. Thus a near-perfect self-predictive island is turned inward upon itself as shown in Figure 9.6. In an active network environment, an executable model can be included within an active packet. When the active packet reaches the target intermediate device, the load model provides virtual input messages to the logical process and the payload of the virtual message is passed to the actual device.

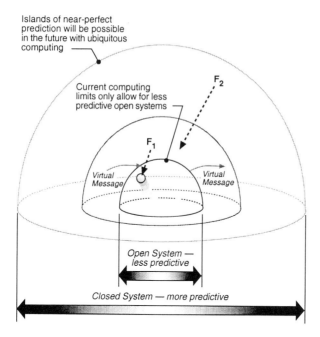

Figure 9.5. Self-predictive islands can improve prediction fidelity by expanding to incorporate more elements.

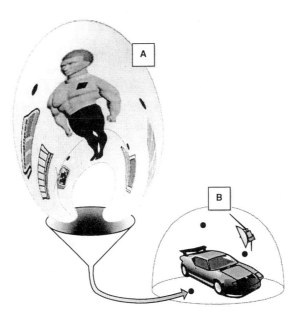

Figure 9.6. Direct communication between A and B is unnecessary as the dynamics of A can be transmitted to B, allowing B to interact with a near-perfect model of A.

Open self-predictive islands will contain inaccuracies in prediction because, by definition, open self-predictive islands include the effects of unknown driving forces upon entities within the of the system. Figure 9.5 shows a force (F_1) acting on the inner system. F_1 is external to the inner system because it is not included within the system itself or in the virtual messages passed into the system. The system could become closed by either enlarging the scope to include the driving forces within the system, as shown in the figure, or by accepting a level of inaccuracy in the system. Thus we can imagine many initial points of near-perfect self-predictive islands, each attempting to improve prediction fidelity by expanding to incorporate more elements. These are the islands of near-perfect self-prediction.

Recursion is a recurring theme in this work. For example, assume that the inner near-perfect self-predictive island in Figure 9.5 is a wireless mobile communications system and F_1 is the weather. Now assume that ubiquitous computing can be used to include weather observation and prediction, for example, computers within planes, cars, spacecraft, etc. The heat from the circuitry of the wireless system, even though negligible, could have an impact on the weather. This is known as the butterfly effect in Chaos Theory. In recent years the study of chaotic nonlinear dynamical systems has led to diverse applications where chaotic motions are described and controlled into some desirable motion. Chaotic systems are sensitive to initial condition. Researchers now realize that this sensitivity can also facilitate control of system motion. For example, in communications, chaotic lasers have been controlled, as have chaotic diode resonator circuits (Aronson et al., 1994, DiBernardo, 1996). Hence, studying the effects of external forces controlling a chaotic system has become a very important goal and should be a subject for research. By allowing for a given tolerance in the amount of error and assuming accuracy in prediction that increases as real time approaches the actual time of an event, this study assumes that a useful near-perfect self-predictive island can be implemented. The Active Virtual Network Management Prediction project attempts to embed predictive capability within an active network using a self-adjusting Time Warp based mechanism for prediction propagation. This self-adjusting property has been found to be useful in prediction and is referred to as autoanaplasis. In addition to autoanaplasis, it is well known that such systems sometimes exhibit super-criticality, faster than critical path execution. However, due to limited and non-ubiquitous computational power in current technology, prediction inaccuracy causes rollbacks to occur. In a world of near-infinite bandwidth and computing power, the cost of a rollback to a "safe" time becomes infinitesimal. This is one of the many new ideas this project will explore involving the relationship between bandwidth, computing power, and prediction. Given near-infinite bandwidth, the system state can be propagated nearly instantaneously. With nearly infinite and ubiquitous computing, driving processes can be developed with near-perfect accuracy. Let us define near-perfect accuracy of our self-adjusting Time Warp based system in the presence of rollback as the characteristic that a predicted state value (V_v) approaches the real value (V_r) as t approaches $GVT(t_1)$ very quickly, where $GVT(t_1)$ is the Global Virtual Time of the system at time t_1. Explicitly, this is, $\forall \varepsilon > 0, \exists \delta > 0$ s.t. $|f(t) - f(GVT(t_1))| < \varepsilon \rightarrow 0 < |GVT(t_1) - t|$ where $f(t) = V_r$ and $f(GVT(t_1)) = V_v$. $f(t)$ is the prediction function. The effect of should not be ignored. These values are described in more detail in Section 5.

9.4.2 Performance Of Near-Perfect Self-Predictive Islands

One focus of study is on the interfaces between systems with various levels of predictive capability. The self-predictive islands formed in Figure 9.3 have various degrees of prediction capability. Our recent theoretical results from the Active Virtual Network Management project indicate that self-predictive islands exhibit high degrees of performance when prediction is accurate, but are brittle when the tolerance for inaccuracy is reached. With respect to network performance as enhanced with Active Virtual Network Management Prediction, systems with little or no prediction capability appear to be ductile, as they are much better able to tolerate prediction inaccuracy, as shown in Figure 9.8. In other words, performance is moderate, but there are no sudden degradations in performance. This compares favorably to a system with a large lookahead and sudden, near catastrophic degradations in performance.

Thus, an obvious question arises as to what is the optimal grouping of predictive components within a system. What happens when the slope shown in Figure 9.8 becomes nearly vertical? The lookahead into the future is tremendously large in some self-predictive islands and smaller in others. If the lookahead is small in a self-predictive island that feeds into a large lookahead system, then large rollbacks are likely to occur. One focus of study is on the interfaces between systems with various levels of predictive capability and the associated "index of refraction" of performance through the interfaces between islands of near-perfect self-prediction.

Brittle behavior of near-perfect self-predictive islands is shown by point D along curve P_h in Figure 9.9. P_h is the performance curve for a high-performance system with brittle characteristics; P_l is a lower-performance system with ductile characteristics. Clearly, the slope from point D along curve P_h is much steeper than that of point E along curve P_l. The steep decline of performance along P_h can be caused by input parameters that exceed a specified tolerance, or by environmental conditions that exceed specified operating boundaries.

Materials Science	Near-Perfect Prediction Systems
Brittle Behavior	Sudden steep decline in performance
Ductile Behavior	Graceful degradation in performance
Stress	Amount parameter exceeds its tolerance
Toughness	System robustness
Hardness	Level of performance within tolerance
Ductility	Level of performance outside of tolerance
Plastic Strain	Degradation from which system cannot recover
Elastic Strain	Degradation from which system can recover
Brittleness	Ratio of hardness over ductility
Deformation	Degradation in performance
Young's Modulus	Amount tolerance is exceeded over degradation

Figure 9.7. Terms Borrowed from Materials Science.

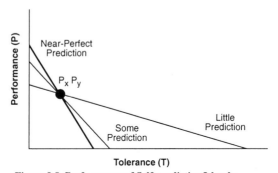

Figure 9.8. Performance of Self-predictive Islands.

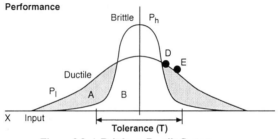

Figure 9.9. A Brittle vs. Ductile System.

Consider a system whose self-predictive islands exhibit various degrees of ductility as defined above. Just as adding impurities to a pure metal causes it to become stronger but more brittle, the addition of more efficient but also more sensitive components to a system, such as a near-perfect self-prediction system, causes the system to increase performance within its operating range, but become less ductile. How do the effects of ductility propagate among the self-predictive islands to influence the ductility of the entire system? Assume the performance response curve is known for each self-predictive island and that the output from one component feeds into the input of the next component as shown in Figure 9.10. The self-predictive islands are labeled S_n and the performance curves as a function of tolerance for error are shown in the illustration immediately above each island. More fundamental research is needed to carry forward this analogy and deliver a theory and models of the relationships among computing, communications, and near-perfect self-prediction.

9.5 SUMMARY

The primary conclusion is that further research is required to understand the nature of entanglement, causality, and the relationship between modeling and communications. For example, Active Network Management Prediction uses a model within a network to enhance the network performance to improve the model's own performance, which thus improves the network's performance thus enhancing the model's performance ad infinitum as shown in Figure 9.4. Furthermore, the Active Virtual Network Management Prediction mechanism uses a Time Warp-like method to ensure causality, yet there is something non-causal about the way Active Virtual Network Management Prediction uses future events to optimize current behavior. This entanglement issue resonates with physicists and those studying the nature of agent autonomy as evident in numerous conferences. Clearly, this needs to be explored in a much deeper manner. Also, formation of islands of near-perfect self-prediction and the need to study the interfaces between those islands was discussed. The idea of wrappers and integration spaces as introduced in (Christopher Landauer and Kirstie L. Bellman, 1996) is likely to provide insight into bringing together complex system components in a self-organizing manner. Another suggestion for the study of predictive interfaces is in a tolerance interaction space (Landauer and Bellman, 1996).

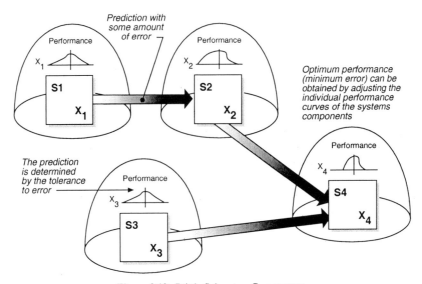

Figure 9.10. Brittle Subsystem Components.

9.6 EXERCISES

1. Can a system predict its own behavior? What are the issues involved?
2. How does AVNMP handle the problem described in the previous problem?
3. If simulation models were nearly perfect representations of real systems, what would be some of the most important implications?
4. Would a model of a system be more efficient to transmit than its individual messages? Under what conditions would this be likely to yield an advantage?
5. Give a simple example of a number in which the transmission of executable code that represents that number is much more efficient than sending the actual number?
6. How will AVNMP behave with imperfect driving process models but high computation speed? What will be the effect as computation speed of individual nodes increases?
7. Use the parameters that affect the performance of AVNMP from the previous chapter's exercise and plot the performance as processing speed increases.

Appendix : AVNMP SNMP MIB

A diagram of the Active Virtual Network Management Prediction SNMP Management Information Base is shown in Figure 9.A.1. This diagram is the authors' interpretation of a Case Diagram, showing the relationship between the primary MIB objects. Many of the MIB objects are for experimental purposes; only the necessary and sufficient SNMP objects based on the authors' experience are included in the Case Diagram. In Figure 9.A.1, the AVNMP process, not shown, can be thought of as being on the top of the figure and the network communication mechanism (not shown) on the bottom of the figure. The vertical arrows illustrate the main path of information flow between the AVNMP process and the underlying network. Lines that cross the main flows indicate counters that accumulate information as each packet transitions between the network and the AVNMP process. Arrows that extend from the main flow are counters where packets are removed from the main flow. The complete AVNMP version 1.1 MIB follows and is included on the CD inmib-avnmp.txt.

AVNMP-MIB **DEFINITIONS** ::= **BEGIN**

IMPORTS
    MODULE-IDENTITY, **OBJECT**-TYPE, experimental,
    Counter32, TimeTicks
    FROM SNMPv2-SMI
    DisplayString
        FROM SNMPv2-TC;

avnmpMIB **MODULE**-IDENTITY   10
    LAST-UPDATED "9801010000Z"
    ORGANIZATION "GE CRD"
    CONTACT-INFO
        "Steve Bush bushsf@crd.ge.com"
    DESCRIPTION
        "Experimental MIB modules for the Active Virtual Network
        Management Prediction (AVNMP) system."
    ::= { experimental active(75) 4 }

--
-- Logical Process Table   20
--

IP **OBJECT IDENTIFIER** ::= { avnmpMIB 1 }

IPTable **OBJECT**-TYPE
        SYNTAX SEQUENCE OF LPEntry
        MAX-ACCESS not-accessible
        STATUS current
        DESCRIPTION   30

        "Table of AVNMP LP information."
   ::= { lP 1 }

lPEntry **OBJECT**-TYPE
    SYNTAX LPEntry
    MAX-ACCESS not-accessible
    STATUS current
    DESCRIPTION
        "Table of AVNMP LP information."
    INDEX { lPIndex }   40
    ::= { lPTable 1 }

LPEntry ::= **SEQUENCE** {
    lPIndex   **INTEGER**,
    lPID   DisplayString,
    lPLVT   **INTEGER**,
    lPQRSize **INTEGER**,
    lPQSSize **INTEGER**,
    lPCausalityRollbacks **INTEGER**,
    lPToleranceRollbacks **INTEGER**,   50
    lPSQSize **INTEGER**,
    lPTolerance **INTEGER**,
    lPGVT   **INTEGER**,
    lPLookAhead **INTEGER**,
    lPGvtUpdate **INTEGER**,
    lPStepSize **INTEGER**,
    lPReal   **INTEGER**,
    lPVirtual **INTEGER**,
    lPNumPkts **INTEGER**,
    lPNumAnti **INTEGER**,   60
    lPPredAcc DisplayString,
    lPPropX   DisplayString,
    lPPropY   DisplayString,
    lPETask   DisplayString,
    lPETrb   DisplayString,
    lPVmRate DisplayString,
    lPReRate DisplayString,
    lPSpeedup DisplayString,
    lPLookahead DisplayString,
    lPNumNoState **INTEGER**,   70
    lPStatePred DisplayString,
    lPPktPred DisplayString,
    lPTdiff   DisplayString,
    lPStateError DisplayString,
    lPUptime TimeTicks
    }

lPIndex **OBJECT**-TYPE
    SYNTAX INTEGER (0..2147483647)
    MAX-ACCESS not-accessible 80
    STATUS current
    DESCRIPTION
        "The LP table index."
    ::= { lPEntry 1 }

lPID **OBJECT**-TYPE
    SYNTAX DisplayString
    MAX-ACCESS read-only
    STATUS current
    DESCRIPTION 90
        "The LP identifier."
    ::= { lPEntry 2 }

lPLVT **OBJECT**-TYPE
    SYNTAX INTEGER (0..2147483647)
    MAX-ACCESS read-only
    STATUS current
    DESCRIPTION
        "This is the LP Local Virtual Time."
    ::= { lPEntry 3 } 100

lPQRSize **OBJECT**-TYPE
    SYNTAX INTEGER (0..2147483647)
    MAX-ACCESS read-only
    STATUS current
    DESCRIPTION
        "This is the LP Receive Queue Size."
    ::= { lPEntry 4 }

lPQSSize **OBJECT**-TYPE 110
    SYNTAX INTEGER (0..2147483647)
    MAX-ACCESS read-only
    STATUS current
    DESCRIPTION
        "This is the LP send queue size."
    ::= { lPEntry 5 }

lPCausalityRollbacks **OBJECT**-TYPE
    SYNTAX INTEGER (0..2147483647)
    MAX-ACCESS read-only 120
    STATUS current
    DESCRIPTION
        "This is the number of rollbacks this LP has suffered."
    ::= { lPEntry 6 }

lPToleranceRollbacks **OBJECT**-TYPE
    SYNTAX INTEGER (0..2147483647)
    MAX-ACCESS read-only
    STATUS current
    DESCRIPTION   130
        "This is the number of rollbacks this LP has suffered."
    ::= { lPEntry 7 }

lPSQSize **OBJECT**-TYPE
    SYNTAX INTEGER (0..2147483647)
    MAX-ACCESS read-only
    STATUS current
    DESCRIPTION
        "This is the LP state queue size."
    ::= { lPEntry 8 }   140

lPTolerance **OBJECT**-TYPE
    SYNTAX INTEGER (0..2147483647)
    MAX-ACCESS read-only
    STATUS current
    DESCRIPTION
        "This is the allowable deviation between process's
    predicted state and the actual state."
    ::= { lPEntry 9 }
                                                                     150

lPGVT **OBJECT**-TYPE
    SYNTAX INTEGER (0..2147483647)
    MAX-ACCESS read-only
    STATUS current
    DESCRIPTION
        "This is this system's notion of Global Virtual Time."
    ::= { lPEntry 10 }

lPLookAhead **OBJECT**-TYPE
    SYNTAX INTEGER (0..2147483647)   160
    MAX-ACCESS read-only
    STATUS current
    DESCRIPTION
        "This is this system's maximum time into which it can
    predict."
    ::= { lPEntry 11 }

lPGvtUpdate **OBJECT**-TYPE
    SYNTAX INTEGER (0..2147483647)
    MAX-ACCESS read-only   170
    STATUS current

DESCRIPTION
     "This is the GVT update rate."
     ::= { lPEntry 12 }

lPStepSize **OBJECT**-TYPE
     SYNTAX INTEGER (0..2147483647)
     MAX-ACCESS read-only
     STATUS current
     DESCRIPTION   180
     "This is the lookahead (Delta) in milliseconds for each
     virtual message as generated from the driving process."
     ::= { lPEntry 13 }

lPReal **OBJECT**-TYPE
     SYNTAX INTEGER (0..2147483647)
     MAX-ACCESS read-only
     STATUS current
     DESCRIPTION
     "This is the total number of real messages received."   190
     ::= { lPEntry 14 }

lPVirtual **OBJECT**-TYPE
     SYNTAX INTEGER (0..2147483647)
     MAX-ACCESS read-only
     STATUS current
     DESCRIPTION
     "This is the total number of virtual messages
     received."
     ::= { lPEntry 15 }   200

lPNumPkts **OBJECT**-TYPE
     SYNTAX INTEGER (0..2147483647)
     MAX-ACCESS read-only
     STATUS current
     DESCRIPTION
     "This is the total number of all AVNMP packets
     received."
     ::= { lPEntry 16 }
                                                                 210
lPNumAnti **OBJECT**-TYPE
     SYNTAX INTEGER (0..2147483647)
     MAX-ACCESS read-only
     STATUS current
     DESCRIPTION
     "This is the total number of Anti-Messages transmitted
     by this Logical Process."
     ::= { lPEntry 17 }

lPPredAcc **OBJECT**-TYPE   220
    SYNTAX DisplayString
    MAX-ACCESS read-only
    STATUS current
    DESCRIPTION
        "This is the prediction accuracy based upon time
    weighted average of the difference between predicted and real
    values."
    ::= { lPEntry 18 }

lPPropX **OBJECT**-TYPE   230
    SYNTAX DisplayString
    MAX-ACCESS read-only
    STATUS current
    DESCRIPTION
        "This is the proportion of out-of-order messages
    received at this Logical Process."
    ::= { lPEntry 19 }

lPPropY **OBJECT**-TYPE
    SYNTAX DisplayString   240
    MAX-ACCESS read-only
    STATUS current
    DESCRIPTION
        "This is the proportion of out-of-tolerance messages
    received at this Logical Process."
    ::= { lPEntry 20 }

lPETask **OBJECT**-TYPE
    SYNTAX DisplayString
    MAX-ACCESS read-only   250
    STATUS current
    DESCRIPTION
        "This is the expected task execution wallclock time for this
    Logical Process."
    ..= { lPEntry 21 }

lPErb **OBJECT**-TYPE
    SYNTAX DisplayString
    MAX-ACCESS read-only
    STATUS current
    DESCRIPTION   260
        "This is the expected wallclock time spent performing a
    rollback for this Logical Process."
    ::= { lPEntry 22 }

lPVmRate **OBJECT**-TYPE
    SYNTAX DisplayString
    MAX-ACCESS read-only
    STATUS current
    DESCRIPTION   270
        "This is the rate at which virtual messages were
    processed by this Logical Process."
    ::= { lPEntry 23 }

lPReRate **OBJECT**-TYPE
    SYNTAX DisplayString
    MAX-ACCESS read-only
    STATUS current
    DESCRIPTION
        "This is the time until next virtual message."   280
    ::= { lPEntry 24 }

lPSpeedup **OBJECT**-TYPE
    SYNTAX DisplayString
    MAX-ACCESS read-only
    STATUS current
    DESCRIPTION
        "This is the speedup, ratio of virtual time to wallclock time,
    of this logical process."
    ::= { lPEntry 25 }   290

lPLookahead **OBJECT**-TYPE
    SYNTAX DisplayString
    MAX-ACCESS read-only
    STATUS current
    DESCRIPTION
        "This is the expected lookahead in milliseconds of this
    Logical Process."
    ::= { lPEntry 26 }

                    300
lPNumNoState **OBJECT**-TYPE
    SYNTAX INTEGER (0..2147483647)
    MAX-ACCESS read-only
    STATUS current
    DESCRIPTION
        "This is the number of times there was no valid state to
    restore when needed by a rollback or when required to check
    prediction accuracy."
    ::= { lPEntry 27 }

                    310
lPStatePred **OBJECT**-TYPE
    SYNTAX DisplayString

MAX-ACCESS read-only
STATUS current
DESCRIPTION
    "This is the cached value of the state at the nearest
time to the current time."
::= { lPEntry 28 }

lPPktPred **OBJECT**-TYPE   320
    SYNTAX DisplayString
MAX-ACCESS read-only
STATUS current
DESCRIPTION
    "This is the predicted value in a virtual message."
::= { lPEntry 29 }

lPTdiff **OBJECT**-TYPE
    SYNTAX DisplayString
MAX-ACCESS read-only   330
STATUS current
DESCRIPTION
    "This is the time difference between a predicted and an
actual value."
::= { lPEntry 30 }

lPStateError **OBJECT**-TYPE
    SYNTAX DisplayString
MAX-ACCESS read-only
STATUS current   340
DESCRIPTION
    "This is the difference between the contents of an application
value and the state value as seen within the virtual message."
::= { lPEntry 31 }

lPUptime **OBJECT**-TYPE
    SYNTAX DisplayString
MAX-ACCESS read-only
STATUS current
DESCRIPTION   350
    "This is the time in milliseconds that AVNMP has been
running on this node."
::= { lPEntry 32 }

END

GLOSSARY

AA Active Application. An Active Application is supported by the Execution Environment on an active network device. The Active Application consists of active packets that support a particular application.

Autoanaplasis Autoanaplasis is the self-adjusting characteristic of streptichrons. One of the virtues of the Active Virtual Network Management Prediction Algorithm is the ability for the predictive system to adjust itself as it operates. This is accomplished in two ways. When real time reaches the time at which a predicted value had been cached, a comparison is made between the real value and the predicted value. If the values differ beyond a given tolerance, then the logical process rolls backward in time. Also, active packets which implement virtual messages adjust, or refine, their predicted values as they travel through the network.

AVNMP Active Virtual Network Management Prediction. An algorithm that allows a communications network to advance beyond the current time in order to determine events before they occur.

C/E Condition Event Network. A C/E network consists of state and transition elements which contain tokens. Tokens reside in state elements. When all state elements leading to a transition element contain a token, several changes take place in the C/E network. First, the tokens are removed from the conditions which triggered the event, the event occurs, and finally tokens are placed in all state outputs from the transition which was triggered. Multiple tokens in a condition and the uniqueness of the tokens are irrelevant in a C/E Net.

CE Clustered Environment. One of the contributions of (Avril and Tropper, 1995) in CTW is an attempt to efficiently control a cluster of LPs on a processor by means of the CE. The CE allows multiple LPs to behave as individual LPs as in the basic time warp algorithm or as a single collective LP.

Channel Channel. An active network channel is a communications link upon which active packets are received. The channel determines the type of active packet and forwards the packet to the proper Execution Environment. Principals use anchored channels to send packets between the execution environment and the underlying communication substrate. Other channels are cut through, meaning that they forward packets through the active node–from an input device to an output device–without being intercepted and processed by an Execution Environment. Channels are in general full-duplex, although a given principal might only send or receive packets on a particular channel.

CMB Chandy-Misra-Bryant. A conservative distributed simulation synchronization technique.

CMIP Common Management Information Protocol. A protocol used by an application process to exchange information and commands for the purpose of managing remote computer and communications resources. Described in (ISO, 1995).

CS Current State. The current value of all information concerning a PP encapsulated by an LP and all the structures associated with the LP.

CTW Clustered Time Warp. CTW is an optimistic distributed simulation mechanism described in (Avril and Tropper, 1995).

EE Execution Environment. The Execution Environment is supported by the Node Operating System on an active network device. The Execution Environment receives active packets and executes any code associated with the active packet.

Fossil In an AVNMP Logical Process, as the Local Virtual Time advances, the state queue is filled with predicted values. As the wallclock advances, the predicted values become actual values. When the wallclock advances beyond the time a predicted value was to occur, the value becomes a fossil because it is no longer a prediction, but an actual event that has happened in the past. Fossils should be removed periodically to avoid excessive use of memory.

FSM Finite State Machine. A five-tuple consisting of a set of states, an input alphabet, an output alphabet, a next-state transition function, and an output function. Used to formally describe the operation of a protocol.

GPS Global Positioning System. A satellite-based positioning service developed and operated by the Department of Defense.

GSV Global Synchronic Distance. The maximum Synchronic Distance in a Petri-Net model of a system.

GVT Global Virtual Time. The largest time beyond which a rollback based system will never rollback.

IETF Internet Engineering Task Force. The main standards organization for the Internet. The IETF is a large open international community of network designers, operators, vendors, and researchers concerned with the evolution of the Internet architecture and the smooth operation of the Internet. It is open to any interested individual.

IPC Inter-Processor Communication. Communication among Unix processes. This may take place via sockets, shared memory, or semaphores.

LP Logical Proces. An LP consists of the PP and additional data structures and instructions which maintain message order and correct operation as a system executes ahead of real time.

LVT Local Virtual Time. The Logical Process contains its notion of time known as Local Virtual Time.

NodeOS Node Operating System. The Node Operating System is the base level operating system for an active network device. The Node Operating System supports the Execution Environments.

MIB Management Information Base. A collection of objects which can be accessed by a network management protocol.

MTW Moving Time Windows. MTW is a distributed simulation algorithm that controls the amount of aggressiveness in the system by means of a moving time window. The trade-off in having no roll-backs in this algorithm is loss of fidelity in the simulation results.

NFT No False Time-stamps. NFT Time Warp assumes that if an incorrect computation produces an incorrect event $(E_{i,T})$, then it must be the case that the correct computation also produces an event $(E_{i,T})$ with the same time-stamp. This simplification makes the analysis in (Ghosh et al., 1993) tractable.

NPSI Near Perfect State Information. The NPSI Adaptive Synchronization Algorithms for PDES are discussed in (Srinivisan and Paul F. Reynolds, 1995b) and (Srinivisan and Paul F. Reynolds, 1995a). The adaptive algorithms use feedback from the simulation itself in order to adapt. The NPSI system requires an overlay system to return feedback information to the LPs. The NPSI Adaptive Synchronization Algorithm examines the system state (or an approximation of the state) calculates an error potential for future error, then translates the error potential into a value which controls the amount of optimism.

NTP Network Time Protocol. A TCP/IP time synchronization mechanism. NTP (Mills, 1985) is not required in VNC on the RDRN because each host in the RDRN network has its own GPS receiver.

PA Perturbation Analysis. The technique of PA allows a great deal more information to be obtained from a single simulation execution than explicitly collected statistics. It is particularly useful for finding the sensitivity information of simulation parameters from the sample path of a single simulation run. It may be an ideal way for VNC to automatically adjust tolerances and provide feedback to driving process(es).Briefly, assume a sample path, (Θ,ξ) from a simulation. Θ is vector of all parameters and ξ is a vector of all random occurrences. $L(\Theta,\xi)$ is the sample performance. $J(\Theta,\xi)$ is the average performance, $E[L(\Theta,\xi)]$. Parameter changes cause perturbations in event timing. Perturbations in event timing propagate to other events. This induces perturbations in L. If perturbations into (Θ,ξ) are small, assume event trace $(\Theta + d\Theta,\xi)$ remains unchanged. Then $dL(\Theta,\xi)/d\Theta$ can be calculated. From this, the gradient of $J(\Theta)$ can be obtained, which provides the sensitivity of performance to parameter changes. PA can be used to adjust tolerances while VNC is executing because event times are readily available in the SQ.

PDES Parallel Discrete Event Simulation. PDES is a class of simulation algorithms which partition a simulation into individual events and synchronizes the time the events are executed on multiple processors such that the real time to execute the simulation is as fast as possible.

PDU Protocol Data Unit. 1. Information that is delivered as a unit among peer entities of a network and that may contain control information, address information, or data. 2. In layered systems, a unit of data that is specified in a protocol of a given layer and that consists of protocol-control information of the given layer and possibly user data of that layer.

P/T Place Transition Net. A P/T Network is exactly like a C/E Net except that a P/T Net allows multiple tokens in a place and multiple tokens may be required to cause a transition to fire.

PIPS Partially Implemented Performance Specification. PIPS is a hybrid simulation and real-time system which is described in (Bagrodia and Shen, 1991). Components of a performance specification for a distributed system are implemented while the remainder of the system is simulated. More components are implemented and tested with the simulated system in an iterative manner until the entire distributed system is implemented.

PP Physical Process. A Physical Process is nothing more than an executing task defined by program code. An example of a PP is the RDRN beam table creation task. The beam table creation task generates a table of complex weights which controls the angle of the radio beams based on position input.

Principal The primary abstraction for accounting and security purposes is the principal. All resource allocation and security decisions are made on a per-principal basis. In other words, a principal is admitted to an active node once it has authenticated itself to the node, and it is allowed to request and use resources.

QR Receive Queue. A queue used in the VNC algorithm to hold incoming messages to a LP. The messages are stored in the queue in order by receive time.

QS Send Queue. A queue used in the VNC algorithm to hold copies of messages which have been sent by a LP. The messages in the QS may be sent as anti-messages if a rollback occurs.

QoS Quality of Service. Quality of Service is defined on an end-to-end basis in terms of the following attributes of the end-to-end ATM connection: Cell Loss Ratio, Cell Transfer Delay, Cell Delay Variation.

RT Real Time. The current wall clock time.

SLP Single Processor Logical Process. Multiple LPs executing on a single processor.

SLW Sliding Lookahead Window. The SLW is used in VNC to limit or throttle the prediction rate of the VNC system. The SLW is defined as the maximum time into the future for which the VNC system may predict events.

SmallState SmallState is a named cache within an active network node's execution environment that allows active packets to store information. This allows packets to leave information behind for other packets to use.

SNMP Simple Network Management Protocol. The Transmission Control Protocol/Internet Protocol (TCP/IP) standard protocol that (a) is used to manage and control IP gateways and the networks to which they are attached,(b) uses IP directly, bypassing the masking effects of TCP error correction,(c) has direct access to IP datagrams on a network that may be operating abnormally, thus requiring management, (d) defines a set of variables that the gateway must store, and (e) specifies that all control operations on the gateway are a side-effect of fetching or storing those data variables, i.e., operations that are analogous to writing commands and reading status. SNMP is described in (Rose, 1991).

SQ State Queue. The SQ is used in VNC as a LP structure to hold saved state information for use in case of a rollback.The SQ is the cache into which pre-computed results are stored.

Streptichron A Streptichron is an active packet facilitating prediction. It is a superset of the virtual message. It can contain self-adjusting model parameters, an executable model, or simple state values.

TR Receive Time. The time a VNC message value is predicted to be valid.

TS Send Time. The LVT that a virtual message has been sent. This value is carried within the header of the message. The TS is used for canceling the effects of false messages.

REFERENCES

Alexander, D.S., Arbaugh, W.A., Hicks, M.W., Kakkar, P., Keromytis, A.D., Moore, J.T., Gunter, C.A., Nettles, S.M., and Smith, J.M. (1998). The SwitchWare Active Network Architecture. *IEEE Network*, 12(3):27--36.

Alexander, D.S., Braden, B., Gunter, C., Jackson, A., Keromytis, A., Minden, G., and Wetherall, D., editors (1997). *Active Network Encapsulation Protocol (ANEP)*. Active Networks Group. Request for Comments: draft

Andre, C., Armand, P., and Boeri, F. (1979). Synchronic Relations and Applications in Parallel Computation. *Digital Processes,* pages 99,113.

Aronson, I., Levine, J., and Tsimring, L. (1994). Controlling Spatio-Temporal Chaos. *Phys. Rev. Lett.*, 72.

Avril, H. (1996). *Clustered Time Warp and Logic Simulation.* PhD thesis, McGill University.

Avril, H. and Tropper, C. (1995). Clustered Time Warp and Logic Simulation.

Bagrodia, R. and Shen, C.-C. (1991). MIDAS: Integrated Design and Simulation of Distributed Systems. *IEEE Transactions on Software Engineering.*

Ball, D. and Hoyt, S. (1990). The Adaptive Time-Warp Concurrency Control Algorithm. In *Proceeding of SCS'90.*

Berry, O. and Jefferson, D. (1985). Critical Path Analysis of Distributed Simulation. In *SCS Multi Conference on Distributed Simulation.*

Boukerche, A. and Tropper, C. (1994). A Static Partitioning and Mapping Algorithm for Conservative Parallel Simulations. In *Proceedings of the 8th Workshop on Parallel and Distributed Simulation,* pages 164,172.

Braden, B., Cerpa, A., Fischer, T., Lindell, B., Kaum, J., and Phillips, G. (2000). Introduction to the ASP Execution Environment. Technical report, USC/ISI. url: http://www.isi.edu/active-signal/ARP/index.html.

Bush, S.F., , Kulkarni, A., Evans, S., and Galup, L. (2000). Active Jitter Control. In *7th International IS\&N Conference, Intelligence in Services and Networks (ISN) '00, Kavouri, Athens, Greece.*

Bush, S.F. (1997). *The Design and Analysis of Virtual Network Configuration for a Wireless Mobile ATM Network.* PhD thesis, University of Kansas.

Bush, S.F. (1999). Active Virtual Network Management Prediction. In *Parallel and Discrete Event Simulation Conference (PADS) '99*.

Bush, S.F. (2000). Islands of Near-Perfect Self-Prediction. In *Proceedings of Vwsim'00: Virtual Worlds and Simulation Conference, WMC'00: 2000 SCS Western Multi-Conference, San Diego, SCS (2000)*.

Bush, S.F. and Barnett, B. (1998). A Security Vulnerability Assessment Technique and Model. Technical Report 98CRD028, General Electric Corporate Research and Development Center.

Bush, S.F., Frost, V.S., and Evans, J.B. (1999). Network Management of Predictive Mobile Networks. *Journal of Network and Systems Management*, 7(2).

Calvert, K. (1998). Architectural Framework for Active Networks (Version 0.9). Active Networks Working Group.

Chandy, K.M. and Misra, J. (1979). Distributed Simulation: A Case Study in Design and Verification of Distributed Programs. *IEEE Transactions on Software Engineering*.

Christopher Landauer and Kirstie L. Bellman (1996). Semiotics of Constructed Complex Systems. In *Intelligent Systems: A Semiotic Perspective Proceedings of the 1996 International Multidisciplinary Conference Volume I: Theoretical Semiotics, Workshop on Intelligence in Constructed Complex Systems*.

daSilva, S., Florissi, D., and Yemini, Y. (1998). Composing Active Services in NetScript. In *Proceedings of the DARPA Active Networks Workshop (Tucson, Arizona)*.

DiBernardo, M. (1996). An Adaptive Approach to the Control and Synchronization of Continuous-Time Chaotic Systems. *Int. J. of Bifurcation and Chaos*, 6(3).

Felderman, R.E. and Kleinrock, L. (1990). An Upper Bound on the Improvement of Asynchronous versus Synchronous Distributed Processing. In *SCS '90*.

Fujimoto, R.M. (1990). Parallel Discrete Event Simulation. *Communications of the ACM*, 33(10):30--53.

Ghosh, K., Fujimoto, R.M., and Schwan, K. (1993). Time Warp Simulation in Time Constrained Systems. In *Proceedings of the 7th Workshop on Parallel and Distributed Simulation*, pages 163,166.

Glazer, D.M. (1993). *Load Balancing Parallel Discrete Event Simulations*. PhD thesis, McGill University.

Glazer, D.M. and Tropper, C. (1993). A Dynamic Load Balancing Algorithm for Time Warp. *Parallel and Distributed Systems*, 4(3):318,327.

Goltz, U. (1987). Synchronic Distance. In *Petri Nets: Central Model and Their Properties. Advances in Petri Nets 1986, Proceedings of an Advanced Course, Bad Honnef. Lecture Notes on Computer Science 254*, pages 338,358. Springer-Verlag.

Goltz, U. and Reisig, W. (1982). Weighted Synchronic Distances. In *Applications and Theory of Petri Nets, Informatik-Fachberichte*, pages 289,300. Springer-Verlag.

Gupta, A., Akyldiz, I.F., and Fujimoto, R.M. (1991). Performance Analysis of Time Warp with Multiple Homogeneous Processors. *IEEE Transactions on Software Engineering*.

Hershey, J. and Bush, S.F. (1999). On Respecting Interdependence Between Queuing Policy and Message Value. Technical Report 99CRD151, General Electric Corporate Research and Development.

Hicks, M., Kakkar, P., Moore, J.T., Gunter, C.A., and Nettles, S. (1999). PLAN: A programming language for active networks. *ACM SIGPLAN Notices*, 34(1):86--93.

Ho, Y.-C. (1992). Perturbation Analysis: Concepts and Algorithms. In *Proceedings of the 1992 Winter Simulation Conference*.

Hoare, C.A.R. (1981). Communicating Sequential Processes. *Communications of the ACM*.

Hofstadter, D.R. (1980). Gödel, Escher, Bach: An Eternal Golden Braid. Vintage Books. ISBN 0-394-74502-7.

Hong, D. and Rappaport, S.S. (1986). Traffic Model and Performance Analysis for Cellular Mobile Radio Telephone Systems with Prioritized and Non prioritized Handoff Procedures. *IEEE Transactions on Vehicular Technology*.

Huber, O.J. and Toutain, L. (1997). Mobile agents in active networks. In *3rd ECOOP Workshop on Mobile Object Systems,* Jyväskylä, Finland.

ISO (1995). Open Systems Interconnection - Management Protocol Specification - Part 2: Common Management Information Protocol.

Jefferson, D.R. and Sowizral, H.A. (1982). Fast Concurrent Simulation Using The Time Warp Mechanism, Part I: Local Control. Technical Report TR-83-204, The Rand Corporation.

Jha, V. and Bagrodia, R.L. (1994). A Unified Framework for Conservative and Optimistic Distributed Simulation. In *Proceedings of the 8th Workshop on Parallel and Distributed Simulation*, pages 12,19.

J. Martinez and Silva, M. (1982). A Simple and Fast Algorithm to Obtain all Invariants of a Generalized Petri Net. In *Proceedings of the Second European Workshop on Application and Theory of Petri Nets*.

Kleinrock, L. (1975). *Queuing Systems Volume I: Theory*. John Wiley and Sons.

Konas, P. and Yew, P.C. (1995). Partitioning for Synchronous Parallel Simulation. In *Proceedings of the 9th Workshop on Parallel and Distributed Simulation*, pages 181,184.

Kulkarni, A.B., Minden, G.J., Hill, R., Wijata, Y., Sheth, S., Pindi, H., Wahhab, F., Gopinath, A., and Nagarajan, A. (1998). Implementation of a Prototype Active Network. In *OPENARCH '98*.

Lamport, L. (1978). Time, Clocks, and the Ordering of Events in a Distributed System. *Communications of the ACM*.

Landauer, C. and Bellman, K.L. (1996). Integration Systems and Interaction Spaces. In *Proceedings of FroCoS'96: The First International Workshop on Frontiers of Combining Systems.*

Lazowaska, E. and Lin, Y.-B. (1990). Determining the Global Virtual Time in a Distributed Simulation. Technical Report 90-01-02, University of Washington.

Legedza, U., Wetherall, D.J., and Guttag, J. (1998). Improving the Performance of Distributed Applications Using Active Networks. Submitted to *IEEE INFOCOM.*

Leong, H.V. and Agrawal, D. (1994). Semantics-based Time Warp Protocols. In *Proceedings of the 7th International Workshop on Distributed Simulation.*

Lin, Y.-B. (1990). Understanding the Limits of Optimistic and Conservative Parallel Simulation. Technical Report UWASH-90-08-02, University of Washington.

Lin, Y.-B. and Lazowska, E.D. (1990). Optimality Considerations of "Time Warp" Parallel Simulation. Technical Report UWASH-89-07-05, University of Washington.

Lipton, R.J. and Mizell, D.W. (1990). Time Warp vs. Chandy-Misra: A Worst-Case Comparison. In *SCS '90.*

Liu, G., Marlevi, A., and Jr., G. Q.M. (1995). A Mobile Virtual-Distributed System Architecture for Supporting Wireless Mobile Computing and Communications. In *Mobicom '95.*

Liu, G.Y. (1996). *The Effectiveness of a Full-Mobility Architecture for Wireless Mobile Computing and Personal Communications.* PhD thesis, Royal Institute of Technology, Stockholm, Sweden.

Liu, G.Y. and Jr., G. Q.M. (1995). A Predictive Mobility Management Algorithm for Wireless Mobile Computing and Communications In *International Conference on Universal Personal Communications (ICUPC)*, pages 268,272.

Lubachevsky, B., Schwatz, A., and Weiss, A. (1989). Rollback Sometimes Works ... if Filtered. In *Proceedings of the 1989 Winter Simulation Conference*, pages 630--639.

Lubachevsky, B.D. (1989). Efficient Distributed Event Driven Simulations of Multiple-Loop Networks. *Communications of the ACM,* 32(1):111--131.

Luenberger, D.G. (1989). *Linear and Nonlinear Programming.* Addison-Wesley.

Ma, S. and Ji, C. (1998). Wavelet models for video time-series. In Jordan, M.I., Kearns, M.J., and Solla, S.A., editors, *Advances in Neural Information Processing Systems*, volume 10. The MIT Press.

Madisetti, V., Walrand, J., and Messerschmitt, D. (1987). MTW: Experimental Results for a Constrained Optimistic Scheduling Paradigm. In *Proc. 1987 Winter Simulation Conference.*

Massechusetts Institute of Technology (1999). http://ana.lcs.mit.edu/darpa/. http://ana.lcs.mit.edu/darpa/.

McAffer, J. (1990). A Unified Distributed Simulation System. In *Proceedings of the 1990 Winter Simulation Conference*, pages 415,422.

Mikami, K., Tamura, H., Sengoku, M., and Yamaguchi, Y. (1993). On a Sufficient Condition for a Matrix to be the Synchronic Distance Matric of a Marked Graph. *IEICE Transactions Fundamentals*, E76-A(10).

Mills, D.L., editor (1985). *Network Time Protocol*. M/A-COM Linkabit.

Murphy, S. (1998). Secure Active Network Prototypes. Active Networks Working Group.

Noble, B.L. and Chamberlain, R.D. (1995). Predicting the Future: Resource Requirements and Predictive Optimism. In *Proceedings of the 9th Workshop on Parallel and Distributed Simulation*, pages 157,164.

Pandit, S.M. and Wu, S.-M. (1983). *Time Series and System Analysis with Applications*. John Wiley and Sons.

Papoulis, A. (1991). *Probability, Random Variables, and Stochastic Processes*. McGraw Hill.

Peterson, J.L. (1981). *Petri Net Theory and the Modeling of Systems*. Prentice-Hall.

Peterson, L. (1998). Node OS and API for Active Networks. Active Networks Working Group.

Rajaei, H., Ayani, R., and Thorelli, L.E. (1993a). The Local Time Warp Approach to Parallel Simulation. In *Proceedings of the 7th Workshop on Parallel and Distributed Simulation,* pages 37--43.

Rajaei, H., Ayani, R., and Thorelli, L.-E. (1993b). The Local Time Warp Approach to Parallel Simulation. In *Proceedings of the 7th Workshop on Parallel and Distributed Simulation*, pages 119,126.

Reisig, W. (1985). *Petri Nets*. Springer-Verlag.

Rose, M.T. (1991). *The Simple Book, An Introduction to the Management of TCP/IP Based Internets*. Prentice Hall.

Samrat Bhattacharjee, Kenneth L. Calvert and Ellen W. Zegura (1998). Reasoning About Active Network Protocols. In *Proceedings of ICNP '98*. Austin, TX.

Seshan, S., Balakrishnan, H., and Katz, R.H. (1996). Handoffs in Cellular Wireless Networks: The Daedalus Implementation Experience. *Kluwer International Journal on Wireless Personal Communications*.

Silva, M. and Colom, J.M. (1988). On the Computation of Structural Synchronic Invariants in P/T Nets. In *Lecture Notes on Computer Science*, volume 340, pages 386,417. Springer-Verlag.

Silva, M. and Murata, T. (1992). B-Fairness and Structural B-Fairness in Petri Net Models of Concurrent Systems. *Journal of Computer and System Sciences*, 44:447--477.

Sokol, L.M., Briscoe, D.P., and Wieland, A.P. (1988). MTW: A Strategy for Scheduling Discrete Simulation Events for Concurrent Execution. In *Proceedings of the 2nd Workshop on Parallel and Distributed Simulation,* pages 34--42.

Sokol, L.M. and Stucky, B.K. (1990). WOLF: A Rollback Algorithm for Optimistic Distributed Simulation Systems. In *Proc. 1990 SCS Multi conference on Distributed Simulation*, pages 169--173.

Srinivisan, S. and Paul F. Reynolds, J. (1995a). Adaptive Algorithms vs. Time Warp: An Analytical Comparison. Technical Report CS-95-20, University of Virginia.

Srinivisan, S. and Paul F. Reynolds, J. (1995b). NPSI Adaptive Synchronization Algorithms for PDES. Technical Report CS-94-44, University of Virginia.

Steinman, J.S. (1992). SPEEDES: A Unified Approach to Parallel Simulation. In *Proceedings of the 6th Workshop on Parallel and Distributed Simulation*, pages 75,84.

Steinman, J.S. (1993). Breathing Time Warp. In *Proceedings of the 7th Workshop on Parallel and Distributed Simulation*, pages 109,118.

Steven Berson and Bob Braden and Livio Riciulli (2000). Introduction to the ABone. http://www.csl.sri.com/activate/.

Tamura, H. and Abe, T. (1996). Obtaining a Marked Graph from a Synchronic Distance Matrix. *Electronics and Communications in Japan*, 79(3).

Tennenhouse, D.L. and Bose, V.G. (1995). SpectrumWare: A Software-Oriented Approach to Wireless Signal Processing. In *Mobicom '95*.

Tennenhouse, D.L., Smith, J.M., Sincoskie, W.D., Wetherall, D.J., and Minden, G.J. (1997). A survey of active network research. *IEEE Communications Magazine,* 35(1):80-86.

Thomas, R., Gilbert, H., and Mazziotto, G. (1988). Influence of the Movement of Mobile Stations on the Performance of the Radio Cellular Network. *Proceedings of the 3rd Nordic Seminar.*

Tinker, P. and Agra, J. (1990). Adaptive Model Prediction Using Time Warp. In *SCS '90.*

Turnbull, J. (1992). A Performance Study of Jefferson's Time Warp Scheme for Distributed Simulation. Master's thesis, Case Western Reserve University.

Voss, K., Genrich, H.J., and Rozenberg, G. (1987). *Concurrency and Nets.* Springer-Verlag. ISBN 0-387-18057-5.

Wetherall, D., Guttag, J., and Tennenhouse, D. (1999). ANTS: Network services without the red tape. *Computer*, 32(4):42-48.

Yemini, Y., Konstantinou, A.V., and Florissi, D. (2000). NESTOR: An architecture for self-management and organization. *IEEE Journal on Selected Areas in Communications.* To appear in the IEEE JSAC special issue on network management (publication 2nd quarter 2000).

Zegura, E. (1998). Composable Services for Active Networks. AN Composable Services Working Group.

INDEX

ABONE, 20

accuracy, 57, 58, 64, 68, 81, 93, 94, 115, 117, 131, 132, 150, 158, 159, 164, 174, 176

Active Application, 12, 20, 21, 56, 175

Active Global Virtual Time Request Packet, 71

Active Names, 26

Active Network Encapsulation Protocol, 12, 20, 37, 60, 69, 181

active network environment, 59, 161

active nodes, 1, 11, 12, 14, 16, 21, 27, 30, 31, 33, 34, 37, 39, 41, 48, 58, 115, 129

Active Quality of Service, 24

Active Virtual Network Management Prediction, 6, 55, 56, 67, 68, 71, 72, 76, 77, 78, 79, 80, 81, 83, 84, 85, 94, 99, 100, 101, 105, 106, 107, 108, 109, 117, 115, 117, 118, 119, 120, 122, 128, 131, 132, 133, 136, 138, 139, 140, 141, 142, 143, 144, 145, 146, 147, 148, 149, 152, 153, 155, 156, 157, 158, 159, 164, 167, 169, 175, 182

Active Virtual Network Management Prediction Algorithm, 67, 68, 78, 109, 115, 117, 118, 119

Active Web Caching, 24, 26

administration, 21, 31

agents, 50, 51, 53, 80, 183

aggregated congestion control, 26

allocation, 13, 18, 36, 41, 46, 47, 49, 51, 81, 117, 178

ANANA, 12

anchored channels, 41, 42, 44, 175

ANEP, 12, 20, 28, 37, 38, 42, 44, 60, 72, 181

anti-message, 72, 87, 100, 112, 153

anti-messages, 63, 69, 72, 74, 89, 100, 132, 153, 178

Anti-toggle, 72, 73

application-specific services, 32

Architectural Framework Working Group, 19

artificial intelligence, 57, 159

asynchronous, 115, 141

attack, 25, 57, 81, 82

Autoanaplasis, 59, 77, 175

best-effort, 43

189